Financing the Agricultural Sector

Westview Special Studies

The concept of Westview Special Studies is a response to the continuing crisis in academic and informational publishing. Library budgets are being diverted from the purchase of books and used for data banks, computers, micromedia, and other methods of information retrieval. Interlibrary loan structures further reduce the edition sizes required to satisfy the needs of the scholarly community. Economic pressures on university presses and the few private scholarly publishing companies have greatly limited the capacity of the industry to properly serve the academic and research communities. As a result, many manuscripts dealing with important subjects, often representing the highest level of scholarship, are no longer economically viable publishing projects--or, if accepted for publication, are typically subject to lead times ranging from one to three years.

Westview Special Studies are our practical solution to the problem. As always, the selection criteria include the importance of the subject, the work's contribution to scholarship, and its insight, originality of thought, and excellence of exposition. We accept manuscripts in camera-ready form, typed, set, or word processed according to specifications laid out in our comprehensive manual, which contains straightforward instructions and sample pages. The responsibility for editing and proofreading lies with the author or sponsoring institution, but our editorial staff is always available to answer questions and provide guidance.

The result is a book printed on acid-free paper and bound in sturdy, library-quality soft covers. We manufacture these books ourselves using equipment that does not require a lengthy make-ready process and that allows us to publish first editions of 500 to 1500 copies and to reprint even smaller quantities as needed. Thus, we can produce Special Studies quickly and can keep even very specialized books in print as long as there is a demand for them.

About the Book and Authors

Farm credit and tax policies have become increasingly important areas of concern for policymakers and agriculturists. Rising levels of debt use among the nation's commercial producers, rising interest rates, and an increased dependence on international commodity markets have contributed to greater income volatility at the farm level, making financial management an important tool for controlling risks.

This book provides a comprehensive overview and analysis of the financial structure of agriculture and the attendant issues and alternatives in farm credit and tax policy. The authors present a detailed financial analysis of the farm sector based on unpublished census data and other primary sources, looking first at the farm sector in the aggregate, then considering subsectors of the farm economy, and finally comparing the farm sector with nonfarm businesses. They also discuss in detail the structure and policy issues related to capital markets for farm firms.

Dean W. Hughes is the director of the Thornton Agricultural Finance Institute and teaches agricultural finance at Texas Tech University. Stephen C. Gabriel is a partner in Farm Sector Economics Associates. Peter J. Barry is professor of agricultural finance at the University of Illinois. Michael D. Boehlje is chairman of the department of agricultural and applied economics at the University of Minnesota.

Financing the Agricultural Sector

Future Challenges and Policy Alternatives

Dean W. Hughes, Stephen C. Gabriel,
Peter J. Barry, and Michael D. Boehlje

Routledge
Taylor & Francis Group
LONDON AND NEW YORK

First published 1986 by Westview Press

Published 2019 by Routledge
52 Vanderbilt Avenue, New York, NY 10017
2 Park Square, Milton Park, Abingdon, Oxon OX14 4RN

Routledge is an imprint of the Taylor & Francis Group, an informa business

Library of Congress Catalog Card Number: 86-50152

ISBN 13: 978-0-367-00526-9(hbk)

Contents

Tables and Figures

Tables

Figures

Acknowledgments

Many people deserve mention for their efforts in producing this book. First, each of the authors would like to acknowledge the comments and criticisms provided by the other authors on the different chapters of the book. Stephen Gabriel deserves special notice for initiating the project.

Production of the final manuscript was done at Texas Tech University where Cindy McClure, Shalane Chamberlain, Barbie Dickensheet, and Cindy Lamborn did the word processing. Nadine Hughes helped with proofreading. Nancy Osborn, Shalane Chamberlain, Curtis Bednarz and Fazal Hag prepared the artwork. George Amols, Susan Price and Curtis Bednarz provided assistance in data collection. Many thanks to all of you.

Dean W. Hughes

1

Introduction

Dean W. Hughes

The objectives of this book are to provide the interested reader with a perspective on past trends and the current financial situation in agriculture, point out some of the arising controversies in farm credit and tax policies, and describe a framework that can be used in analyzing the costs and benefits of different policy prescriptions. Material is presented as a set of essays on topics related to agricultural finance which can be read in sequence or as individual manuscripts.

Many farmers are currently facing significant financial difficulties. For the first time in over forty years, farm debt accounts for over 20 percent of the value of farm assets. And, highly leveraged farmers are going out of business at a record rate each year.

Historically farmers have gone through cycles of growth and decline in their use of credit. Such cycles take a long time to complete, usually lasting more than a generation. Farmers became wary of the use of debt after the large number of farm foreclosures associated with the years of the Great Depression. Farm debt use remained low until the 1950s, when rapid growth in farm credit began. During the 1970s, farm debt grew at an average annual rate of 12 percent and reached about 16 percent of the value of farm assets in 1980. By the end of 1983, farm debt was 21 percent of the value of farm assets.

Growth in interest expenses exhibited about the same trend as the level of debt until 1979. Rapid increases in interest rates since then have caused interest expenses to soar. In 1979, farm interest expenses accounted for 11 percent of all farm production expenses, but by 1983 they were over 16 percent of expenses.

Many significant changes, besides the growth in farm

debt, have combined to reduce the historic financial isolation of the farm sector over the last two decades. Rapid growth of agricultural exports coupled with the switch to floating exchange rates in the early 1970s increased the sensitivity of agricultural commodity prices to domestic and foreign macroeconomic policies. Banking deregulation has served to integrate rural financial markets into the national money markets, making the interest rates farmers pay on their loans and receive on nonfarm investments far more sensitive to national economic conditions.

There is little doubt that farming today must compete as a business against all other businesses in the national and international capital markets. Without government intervention, every dollar that a U.S. farmer borrows must show as high a rate of return as dollars borrowed by General Motors, American Telephone and Telegraph, and computer firms in the Silicon Valley. With the growing linkages in worldwide financial markets, U.S. farm loans must also be bid away from Volkswagen, Nissan, and Sony. This is substantially different from the past, when pools of rural savings were dedicated to use as farm loans, and rural interest rates reflected only the conditions of local markets.

In the future, farms will be faced with an economic environment very much like the conditions faced by other businesses. The rural economy will probably continue to be more closely linked to the national economy than it was in the past, and farmers will be exposed to more of the economic risks faced by other businesses. Farm interest rates will respond more quickly and more dramatically to changes in national economic polices. The demand for farmers' production will also be subject to changes that farmers will not be able to control. Moreover, it is likely that continuing declines in farm numbers and population will lead to reduced political influence and a less farm-oriented food and fiber policy.

Farming is not a business like any other, however. There are at least two important differences between farming and other production activities: (1) everyone depends on food for life, and (2) farming is conducted in an uncontrollable environment. The first difference makes it clear that the public has serious and legitimate concerns about the efficiency and reliability of the agricultural industry. The second difference implies that farmers face larger risks than other producers, and without public intervention, such risks lead to the underproduction and the overpricing of food. These two differences can form the basis of a strong argument in favor of government farm policies. Questions remain, however, about what policies are best for all concerned: farmers, agribusinesses, consumers, and taxpayers.

Over the long run, the way to expand output in agriculture, or any other industry, is to increase the risk adjusted, after-tax return for investing in the industry relative to the returns available in other industries. Resources eventually flow to industries where investors can generate the highest return on their dollars after risks and taxes have been discounted. Higher levels of resources over time lead to more production and lower prices for consumers.

There are three general ways of increasing the return to an industry. Net incomes can be supported at higher levels than would occur without intervention, which is part of what farm commodity programs and special farm tax provisions have been designed to accomplish. Risks can be lowered by providing insurance, which is the goal of government crop insurance and one interpretation of the objectives of government disaster relief payments and loan programs. Finally, the cost of borrowing capital for investing in the industry can be lowered, with an outstanding example being subsidized interest rates on some government loans to farmers.

At first glance, farm credit and tax policies offer attractive alternatives to commodity policies that can, with relative ease, be applied to all food and fiber producers. Such policies could reduce the need for the government to be involved in the numerous decisions required by commodity policies. Credit and tax policies, however, suffer from a lack of short-run control that many policymakers think is necessary. Moreover, it is difficult to design credit and tax policies that benefit specific, deserving groups of producers. Adjusting interest rates on farm loans or changing farm income tax rates takes a long time to have an impact on farm production. Credit policies also run the risk of creating disturbances in financial markets and may cause farmers to offset financial advantages by accepting additional business risks. These conditions and trade-offs are reflected throughout the book. It is, therefore, unlikely that credit and tax policies will ever completely remove the need for more immediately felt farm programs. They can, however, act as long-term regulators of farm output and reduce the precision needed to implement many of the current farm commodity policies.

Background material is presented in the first few chapters of the book. Chapter 2 describes historic trends in the financial condition of the farm sector as a whole. Chapter 3 extends the description of the current financial problems of farmers by disaggregating financial information by size of farm, region of the country and type of commodity produced. Chapter 4 describes the current markets for farm capital, both equity and debt. Chapter 5 links the beginning chapters of the book to the rest of the book that deals with current and future financial and tax policy issues. It does so by discussing the methods used to project future financial conditions and then presents some possible future financial environments for farm capital markets. Chapter 6 discusses current and future policy choices to be made regarding private farm capital markets. Chapter 7 provides a similar discussion on the future course of more direct intervention by the government in farm finance. Chapter 8 describes the impacts of the many aspects of tax policy on the production, marketing and financing decisions made by farmers. Chapter 9 provides a framework for identifying the likely responses of farmers to government decisions. An analysis of currently debated responses to farm credit problems is also presented using this framework. Finally, Chapter 10 provides a brief summary of the book and the conclusions reached.

2

A Financial Analysis of the Farm Sector: Sector Trends

Stephen C. Gabriel

The purpose of a financial analysis of the farm sector is to obtain a multi-dimensional evaluation of the financial status of the sector vis a vis some objective norm, for the sector itself, or performance levels in other industries, or a trend over time suggesting the extent to which progress has been made. In this chapter we will consider trends in sector income, wealth, debt and cash flow since 1960, as well as various key financial ratios.

First, the income characteristics of the sector will be analyzed in terms of their level, trend, and stability. Accumulation of farm wealth is also an objective of farm managers. These are the two primary aggregate performance measures and together determine rates of return on investment, borrowing capacity, and risk. There is a simultaneous relationship between debt and income and wealth. Higher debt increases interest expenses, thereby reducing net income. At the same time, lower income makes debt less attractive. The converse is also true. Changes in the level of debt have a direct impact on wealth since wealth is defined to be the difference between the value of assets and debt. Higher debt reduces wealth, other things the same. Decreased wealth leads to a lower capacity to borrow. While this description is over simplified, it does demonstrate some of the inter-relationships involved.

Farm and Off-Farm Income Over Time

In this section several measures of farm income are to be considered--net farm income, net cash income, and off-farm income, all measured in real rather than nominal terms. Net farm income measures the value of the farm product produced in a given year after deducting the cost of producing it. Costs of production include noncash items such as depreciation. Income, on the other hand, includes noncash income such as the imputed net rental value of operator dwellings. Hence, net farm income does not reflect the cash income earned in a particular year but rather the income associated with a particular crop. Sales from accumulated inventories are not reflected in net farm income.

Net cash income, on the other hand, measures, as the name implies, the net cash income earned in a given year, and more accurately reflects the cash flow position of the sector during the year. Although net farm income and net cash income move together on trend, from year to year their trends can and have diverged--primarily due to large swings in crop inventories.

Off-farm income again is self-explanatory. It represents any income earned off the farm. Some typical sources of off-farm income include wages, interest, dividends, and returns from non-farm businesses.

Trends in real farm income (1972 dollars) have not been favorable for the past twenty years (Figure 2.1). Real net farm income has been on a downward trend since 1960, the beginning of the time period considered in this analysis. During the sixties the slope of the trend line was only slightly negative. The seventies, however, ushered in a period of extreme income variability and since its peak in 1973, there has been a fairly steep decline in real net farm income.

Real net cash income, which does not include non-money income, value of inventory adjustment, and noncash expenses, has performed somewhat better. Its trend has been upward from 1960 to 1983, overall. But if this period is segmented into two subperiods, 1960-69 and 1970-83, the slope of the trend line is positive during the sixties and negative thereafter. Ignoring the large surge in net cash income in 1973 and considering the trend for 1975-83, the slope is still negative. Hence, taken in the aggregate, the sector as a whole has generated less income in the early eighties, in constant dollars, than it did at the beginning of the previous decade.

The off-farm income of farm families has had a positive and stabilizing effect on total sector income.

After rising steadily throughout the sixties and into the

FIGURE 2.1: FARM INCOME TRENDS

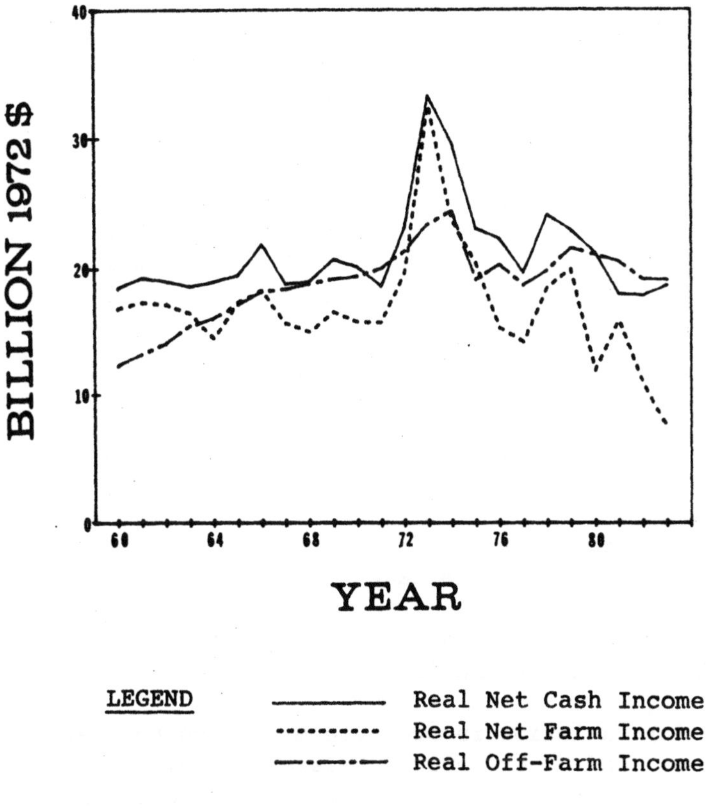

seventies, peaking in 1974, real off-farm income dropped sharply in 1975 and has edged upward only slightly on trend since. The added stability offered by off-farm income, however, has not been evenly distributed among farmers, as will be discussed in the next chapter.

Farm income reached high levels of variability in the middle 1970s. The degree of income variability since 1960 is measured using a moving five-year coefficient of variation (CV) of real net cash income (Table 2.1). The CV in the sixties averaged 4.4 percent and jumped to 14.3 percent during the seventies. Even after removing the 1973 observation, the CV in the seventies averaged 11.5 percent. Hence, the relative volatility of real net cash income increased dramatically in the seventies and has remained high, tapering off somewhat since 1978.

TABLE 2.1: INCOME VARIABILITY IN U.S. AGRICULTURE

Year	CV[1]	Year	CV[1]
1960	5.3	1972	8.0
1961	4.8	1973	23.2
1962	4.0	1974	22.9
1963	2.4	1975	20.7
1964	1.5	1976	17.0
1965	1.6	1977	20.2
1966	6.1	1978	14.3
1967	6.3	1979	7.0
1968	6.0	1980	7.2
1969	5.9	1981	10.6
1970	5.7	1982	10.4
1971	4.1	1983	10.4

[1] Five year moving coefficient of variation of net cash income in percentages.

Miller, et al. summarize many of the factors which led to greater instability in the 1970s.

> (a) Variation in crop yields caused by weather, disease, and other natural hazards was amplified by expanding world production onto semi-arid lands in the 1970's, (b) increased trade among countries with variable production, (c) the adoption of floating exchange rates in 1972, that allow world economic conditions to impact directly on U.S. grain prices, and (d) the use of trade policies by both grain importing and exporting countries to insulate themselves from world market variability and to "export" the impact of their own crop production variability to the world market.
>
> The potential for world market variability, in the 1980's and 1990's is greater than in the 1970s. With an increasing number of domestic markets insulated from world prices, an increasing adjustment burden must be met by the residual world market

and the domestic U.S. market.

The data suggest that the sharp increase in income volatility occurred suddenly and without warning. In the early part of the decade this development was not clear. In fact, it appeared that income simply reached new heights, warranting higher levels of investment and justifying greater financial leverage. Indeed, these expectations were widely held as reflected in rapidly rising land values during that time. It appears that this increase in business risk was followed by higher financial risks in the later 1970s and early 1980s due to the effects of high financial leverage for many farmers, and high, volatile interest rates. Thus, business and financial risks have combined to significantly increase the total risk position of the farm sector.

Income Per Farm

Since 1960 the number of farmers in the sector has declined almost 40 percent. Hence, the farm income "pie" is being divided among fewer farmers. This fact is reflected in the 1960-70 trends in real income per farm (Figure 2.2). For example, during this period real net cash income per farm rose at a compound annual rate of about 4 percent. Real off-farm income per farm rose at an annual rate of 7.7 percent, becoming an increasingly important source of income for many farm families. In the aggregate, off-farm income contributes over 70 percent of total farm family income, up from 43 percent in 1960. However, both real net cash income and net farm income per farm peaked in 1973 and have been declining on trend ever since.

Off-farm income per farm has continued to rise on trend, however, at a much reduced rate of growth. Trends in both income and farm numbers explain the trend in income per farm. Interestingly, the rate of decline in the number of farms dropped in the seventies as farm income waned. No doubt the rate of decline in the number of farms moderated due to the brief surge in farm income in 1973 and 1974 and the subsequent farmland boom which resulted in considerable capital gains for many farmers and investors.

Cost of Production

The cost of producing agricultural products has been rising rapidly over the years. In fact, the growth in

FIGURE 2.2: REAL AVERAGE FARM INCOME

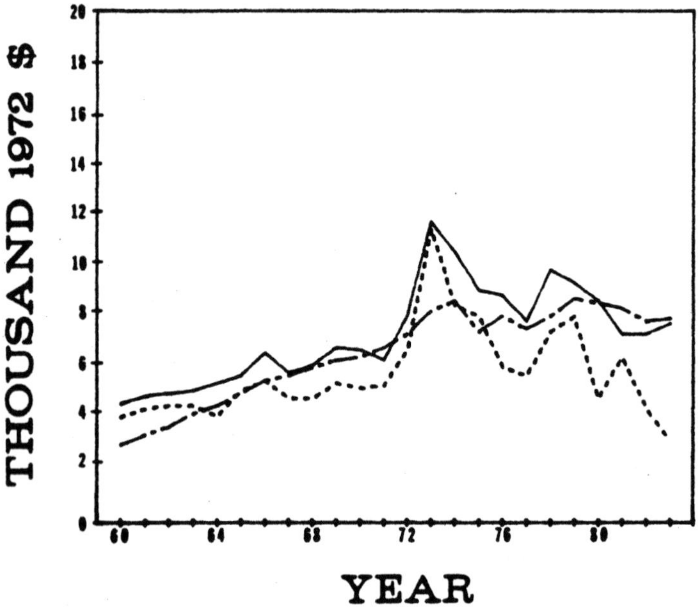

LEGEND
———— Average Real Net Cash Income
·········· Average Real Net Farm Income
—·—·— Average Real Off-Farm Income

production expenses has accelerated since 1960--expenses grew in real terms at a compound annual rate of 2 percent from 1960 to 1970 and a rate of almost 3 percent since 1970. The fastest growing components of production expenses since 1970 have been interest and energy.

Since 1960 production costs have grown faster than cash receipts. This trend is primarily a result of changing price relationships rather than technical productivity. In fact, the index of farm technical productivity grew at a compound annual rate of 2.4

percent from 1970 to 1982, compared with 1.1 percent during the sixties.

Farm production expenses have been consuming an increasing share of total cash receipts as indicated by the gross ratio (production expenses divided by cash receipts) (Figure 2.3). In 1960, production expenses were about 80 percent of receipts. The ratio increased gradually to 89 percent in 1967 and leveled off through 1971. The farm income boom in 1972-74, pulled the ratio down to 74 percent in 1973 and 77 percent in 1974. However, production expenses' claim on cash receipts has been rising on trend, reaching 98 percent in 1983.

FIGURE 2.3: THE GROSS RATIO AND THE DEBT BURDEN

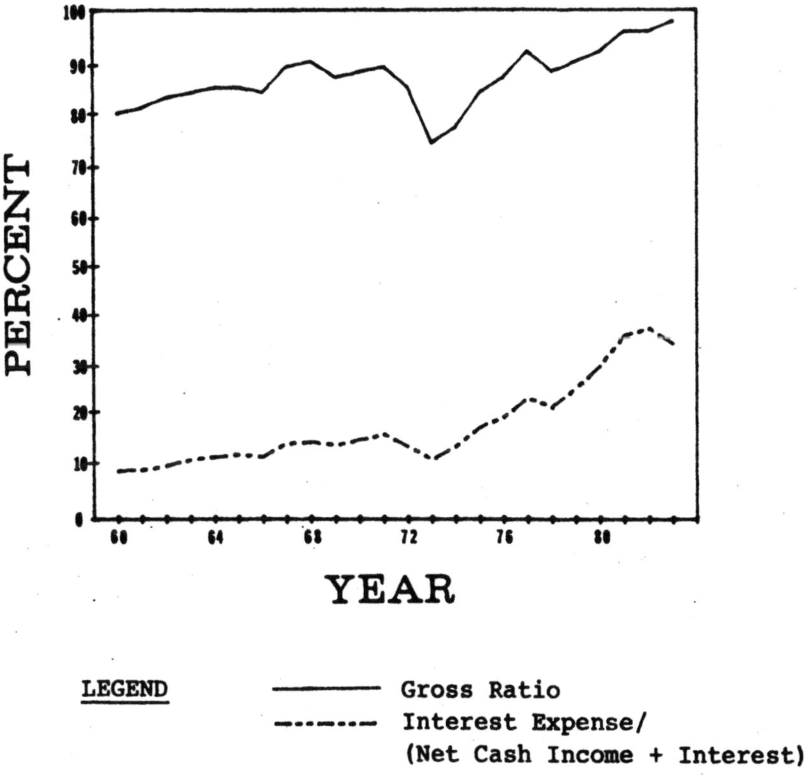

The implication of this trend is that the gross margin (cash receipts less production expenses) has become more volatile for the farm sector as a whole. That is, any given percent change in cash receipts will result in an increasingly greater percent change in gross margin, all other things the same. Hence, the cost structure of agriculture has introduced an additional source of risk which serves to magnify the risk originating in the marketing or production process.

Debt Burden

The debt burden, measured by the ratio of interest expenses to net cash income before deducting interest, rose slightly during the sixties. This period was characterized by great stability in most key financial variables. Interest rates on farm loans ranged from 5 to 7 percent. Farm debt grew modestly compared to the seventies--at a compound annual rate of 7.9 percent. Farm income was stable in real terms. These factors led to an upward drift in the debt burden ratio, rising from 9.6 percent in 1960 to 15.6 percent in 1970.

The debt burden ratio, remaining stable during the next two years, dropped sharply in 1973 with the surge in income that year and then began to soar. The driving force behind the rapid climb in the debt burden was an accelerating growth in farm debt, increasing at an annual rate of 9 percent in the first half of the seventies and 15 percent during the second half. Expectations of higher future incomes, a booming farmland market, and high inflation encouraged many farmers to use financial leverage to take advantage of the boom. Debt financed much of the growth in capital formation during this period.

Since 1980, the growth in farm debt moderated substantially. However, lower incomes and higher interest rates have pushed the debt burden still higher. The index of the debt burden rose from 11.7 percent in 1973 to 37.6 percent in 1982.

The principal implication of this trend in the debt burden is greater volatility in net income. Financial values, trends in real estate values mirror those of the total assets, peaking in 1982 and declining since then.

The relative importance of farm real estate has grown from about 65 percent of total farm assets in 1960 to about 75 percent. Most other asset categories have declined as a percent of the total. However, some types of assets (machinery and crops in storage) have grown in relative importance since the softening of the farmland market in the early eighties.

FIGURE 2.4: REAL FARM ASSETS

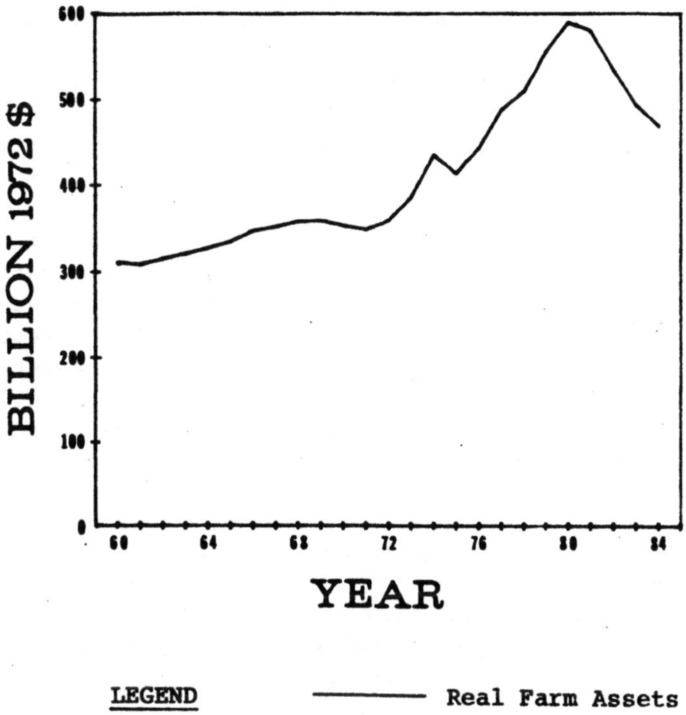

Farm Sector Wealth

The most widely used measure of wealth is equity, or the difference between total assets and total debt. Hence, farm equity measures what a farmer would have available if he were to liquidate all of his assets and pay off all of his debts. Comparisons of equity levels over time require that the dollars measured be comparable leverage has the effect of magnifying the percent changes in net income relative to net income before deducting interest, making highly leveraged farmers even more vulnerable to unstable commodity markets.

Farm Sector Balance Sheet Over Time

Trends in the balance sheet of the farming sector are dominated by real estate assets. Comprising about 75 percent of total farm assets, changes in real estate side of the balance sheet, but the liability side as well. Changes in the value of farm real estate translate directly into changes in farm equity. Real estate values also affect farmers' borrowing capacity and the total return to farmers' investment, both of which influence the growth of farm debt.

The total value of farm assets grew at about a 4 percent rate during the sixties to $315 billion at the end of the decade. In real terms farm assets grew at a rate of 1.4 percent per year. The farm boom of the seventies, however, sparked an accelerated rate of growth (12.3 percent in nominal terms and 5.3 percent in real terms) in farm asset values that brought total assets past the trillion dollar mark. Total farm assets sreached $1015.3 billion in 1980, peaked at $1111.1 billion in 1982, and have consistently declined since then. The 1985 value of farm assets has been estimated at $955.8 billion.

All farm assets grew more rapidly during the seventies than the previous decade due, in part, to a much greater rate of inflation int he seventies. However, inflation does not explain all of the difference in growth rates. In both decades, investment in farm assets was a good hedge against inflation. In the sixties the GNP deflator rose at a compound annual rate of about 3 percent. This same measure of the price level rose at a 7 percent rate over the period of the seventies. After adjusting for inflation, arm asset values grew at an annual rate of 1.2 percent and 5.4 percent in the sixties and seventies, respectively (Figure 2.4).

The value of farm real estate grew at a more rapid annual rate than all the other asset components since 1960--4.6 percent in the sixties and 13.4 percent in the seventies. Being the driving force behind total asset in value. Hence, analyzing the trend in real farm equity allows one to assess the sector's progress in accumulating wealth over time.

During the sixties real equity in the farm sector grew only slightly, at a compound annual rate of less than 1 percent (Figure 2.5). However, the farm boom of the seventies sparked a surge in real farm wealth that began in 1972 and continued at a 5.8 percent annual rate through 1979. To the chagrin of those who invested heavily in farm assets late in the seventies, farm sector real wealth has been declining through 1985.

FIGURE 2.5: REAL FARM EQUITY

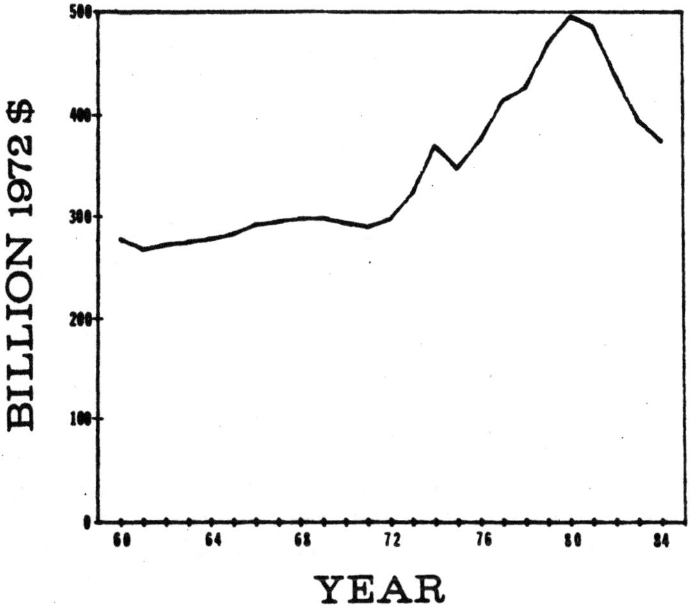

LEGEND ——— Real Farm Equity

Trends in farm sector wealth are important as indicators of financial performance if it is assumed that farmers have as one of their business objectives the accumulation of wealth. Farm wealth not only represents a source of future income, it is a measure of solvency, or risk bearing capacity. As such, it contributes to the total borrowing capacity of the farm or the sector.

Farm Sector Liquidity

Liquidity refers to the extent to which one can convert assets into cash, the speed with which it can be accomplished, and the price discount required to do so. Clearly, one can find a price at which any asset will sell in a very short period of time. Usually, the less active the secondary market, the greater the price discount necessary to sell the asset quickly. Thus, an asset with an inactive secondary market is considered to be illiquid. Farmland is a good example of an illiquid asset, with only two to three percent of all farmland changing hands in a year. A very liquid asset would be a common stock which is traded on a major exchange. Other highly liquid farm assets are crop and livestock inventories. Markets for these assets are well developed and quite active. Sales of such assets can generate cash quickly.

The upward trend in farm real estate's importance in the balance sheet suggests a less liquid asset structure for agriculture. This is particularly so when one considers the decline in financial assets, the most liquid of assets, as a percent of the total. However, rising real estate values in the seventies contributed to farm liquidity in a way that shows up only indirectly on the balance sheet. As discussed in the previous section, the surge in farm real estate values which began in 1972 resulted in a rapid increase in real wealth for the farm sector. Higher levels of farm wealth meant greater borrowing capacity, an important source of liquidity. The degree of liquidity possessed by one's borrowing capacity depends on the certainty with which one can borrow a known quantity and the speed with which the loan can be made. These factors are influenced not only by one's own financial condition, but by external factors as well. Such factors include lender credit policies, macroeconomic developments, and financial and institutional developments which affect the financial intermediation process locally or nationally. It is likely that the rise in real estate values which occurred in the seventies generated more liquidity in borrowing capacity than the sector lost as real estate assets increased as a percentage of total farm assets. However, borrowing capacity is a difficult "asset" to manage and can quickly erode as has been demonstrated in the early eighties.

Farm Sector Debt

An historical discussion of farm debt since 1960 can be segmented into three discrete periods--pre-boom (1960-71), boom (1971-80) and post-boom (1980-84). During the pre-boom period real farm debt rose at an annual rate of 4.4 percent (Figure 2.6). Although nominal farm debt grew at a rate of 13.2 percent during the farm boom compared to 7.4 percent in the pre-boom period, real farm debt grew at a rate of only 5.9 percent, still a significant increase over the pre-boom period.

FIGURE 2.6: REAL FARM DEBT

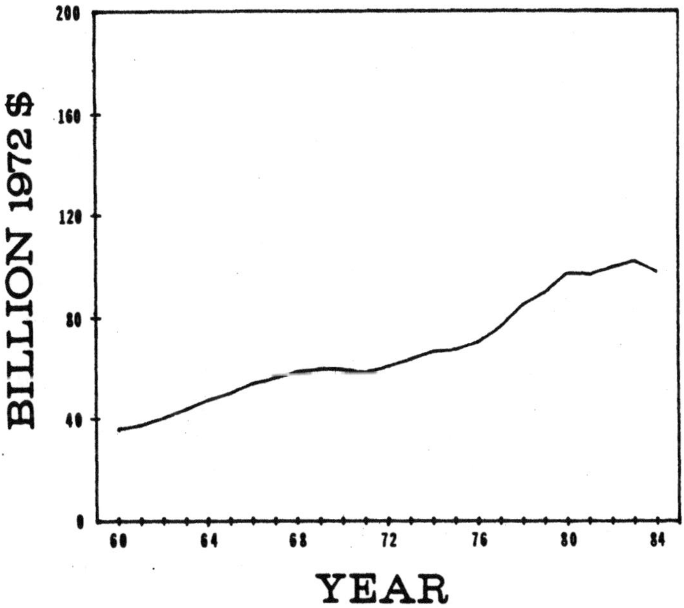

LEGEND ——— Real Farm Debt

The growth in farm debt during the seventies was encouraged by relatively low real interest rates; high unanticipated inflation, which allowed repayment of debt with cheaper dollars; higher incomes and expectations of higher incomes in the future; and rapidly rising farmland values. A favorable relationship between the total return on farm assets (income and capital gains returns) and interest rates suggests that for many farmers the use of financial leverage was an effective strategy for wealth accumulation. Apparently, many farm lenders were aggressive during the boom in agriculture, making loans when cash flow would not support the loan payments. But, rapidly rising farmland values provided security and a basis for refinancing in the future.

It appears, in retrospect, that aggressive financial management strategies and liberal lending practices on the part of lenders during the boom period led to many of the financial difficulties being experienced in the eighties. The eighties ushered in a period of high interest rates, low farm income compared to the seventies, and declining farmland values. The conditions just described led to a drop in borrowing capacity and the liquidity associated with it. The demand for credit also declined for the same reasons. The result was a drop in the real growth in farm debt to 1.6 percent per year.

Sources and Uses of Funds in the Farm Sector

Sources of funds (SOF) in the farm sector reflect trends in farm income. In real terms, SOF exhibited considerable stability in the sixties, rising on trend to a compound annual rate of 2.2 percent. Real SOF peaked in 1973 but have been declining on trend thereafter.

The net flow of loans as a source of funds grew considerably in importance in the middle to late seventies, increasing from an average 9.8 percent of total sources of funds in the sixties to an average 16.2 in the seventies. This source of funds grew to contribute as much as 26 percent of the total in 1977. In 1982, however, the net flow of loans as a percent of total sources of funds dropped sharply to 16.7 percent from 21.5 percent in 1981. It dropped further in 1983 to -2 percent as a result of a net reduction in total farm debt. The growth in the use of loans reflects earlier discussions of the farm boom in the seventies and the rapid growth in farm debt and subsequent slow-down in the eighties.

Uses of funds are divided into two broad categories, gross capital expenditures and a residual item called other cash uses which includes personal consumption, nonfarm investments, and real estate purchases from discontinuing proprietors. Although gross capital expenditures grew rapidly since 1960, peaking in 1979 at $21 billion, capital expenditures as a percent of total cash uses has remained relatively stable in the 16 to 24 percent range. The ratio averaged 20.5 percent in the sixties and 20.1 percent in the seventies and early eighties. The percentage of funds used to purchase capital equipment reached a low of 15.8 percent in 1973. The ratio increased steadily until 1978 in response to relatively favorable farm income. Since peaking at 23.9 percent in 1978, however, capital expenditures have been capturing a declining share of available funds. The reversal of this trend reflects the difficult financial conditions which have emerged in the eighties. High interest rates, relatively low farm income, and weak farmland markets forced many farmers to restructure their balance sheets and postpone capital purchases while reducing their debt burden.

Analytical Ratios

Several key ratios can be derived using data from income, balance sheet, and cash flow statements. Generally, such ratios assist the analyst in assessing the profitability and risk of the sector as a whole.

Farm Sector Profitability

Rates of Return in the Farm Sector Rates of return in farming are derived from two sources, the income generated by farm assets and real capital gains associated with changes in the constant dollar value of farm assets and debt. Both the rates of return to farm assets and farm equity will be considered in this section. The rate of return on farm assets measures the profitability of those assets after paying such factors as operators' labor and management. The rate of return on equity takes into account the cost of financing and provides a rate of return to the owners' investment.

The rate of income return to farm assets in the seventies was higher, on average, than that of the sixties, 5.0 percent versus 3.8 percent (Table 2.2). Again the boom years of the early to mid-seventies boosted the decade average considerably with returns as high as 9.7 percent in 1973. The early eighties, however, show an apparent reversion to the lower rates of income return of the sixties. Substantial capital gains accompanied the higher incomes of the seventies bringing the total rate of return to farm assets to extraordinary levels--15.0 percent, 22.6 percent, 14.0 percent, and 14.1 percent in 1972, 1973, 1975, 1976, respectively. However, just as asset values rose in response to expectations of higher incomes in the seventies, producing substantial real capital gains, revised expectations of lower incomes in the early eighties resulted in capital losses which wiped out any positive income returns, leaving a negative total rate of return to farm assets in 1981 and 1982.

Returns to equity reflect developments in farm sector income and asset values, but also incorporate the cost of borrowing. Hence, the rate of return to equity provides a measure of the profitability of investment in agricultural assets. The pattern of rates of return to equity is similar to that of assets. As with returns to assets, capital gains provide a very important source of total returns to equity. In fact, in some years during the seventies capital gains were two and three time that of income returns. Since 1981, however, capital losses combined with low income have resulted in negative rates of return to equity for the farm sector as a whole. As a result net investment in agriculture has been negative since 1981.

Farm Sector Risk

Debt Burden and Gross Ratios Two key ratios in the section on farm income--the ratio of total production expenses to cash receipt (gross ratio) and the ratio of interest expenses to net cash income before deducting interest (debt burden ratio) have been discussed. Both of these ratios have risen to high levels in recent years, indicating a more volatile residual income stream for the sector. Hence, agriculture has developed a riskier financial structure over the years, allowing little room for management error.

Farm Sector Debt-to-Asset Ratio The debt-to-asset ratio is useful in assessing the financial structure of the sector. It indicates the importance of debt in financing agricultural investment and is a measure of the solvency of the sector. Analysis of trends in the debt-to-asset ratio, therefore, can be useful in evaluating financial conditions in the farm sector.

TABLE 2.2: RATES OF RETURN IN THE FARM SECTOR, 1960-83[1]

	Percentage Rates of Return on Assets			Percentage Rates of Return on Equity		
Year	Income	Capital Gains	Total	Income	Capital Gains	Total
1960	3.1	0.1	3.2	2.7	0.2	2.9
1961	3.7	3.9	7.6	3.3	4.6	7.9
1962	3.8	3.1	6.9	3.5	3.6	7.2
1963	3.7	2.4	6.1	3.3	2.9	6.2
1964	3.1	3.6	6.7	2.7	4.3	7.0
1965	4.6	5.8	10.4	4.4	6.9	11.3
1966	4.8	3.3	8.1	4.6	4.0	8.6
1967	3.8	2.1	5.9	3.2	2.5	5.8
1968	3.6	1.5	5.1	3.0	1.9	4.9
1969	4.1	-0.3	3.8	3.7	-0.3	3.3
1970	4.1	-0.3	3.8	3.5	-0.3	3.2
1971	4.1	4.7	8.8	3.4	5.6	9.1
1972	5.5	9.5	15.0	5.3	11.6	16.9
1973	9.7	12.9	22.6	10.2	15.7	25.8
1974	6.0	0.5	6.5	5.7	0.6	6.3
1975	5.4	8.6	14.0	4.9	10.3	15.2
1976	3.7	11.6	15.3	2.8	13.9	16.7
1977	3.1	4.7	7.8	2.1	5.6	7.7
1978	4.2	9.9	14.1	3.3	12.0	15.3
1979	4.4	3.0	7.4	3.4	3.6	7.0
1980	2.8	-0.5	2.3	1.3	-0.6	0.7
1981	3.7	-7.6	-3.9	2.1	-9.2	-7.1
1982	3.2	-6.3	-3.1	1.3	-7.8	-6.5
1983	2.6	-2.4	0.2	0.5	-3.1	-2.6

[1]Source: U.S. Department of Agriculture

FIGURE 2.7 : INDICES OF THE DEBT BURDEN

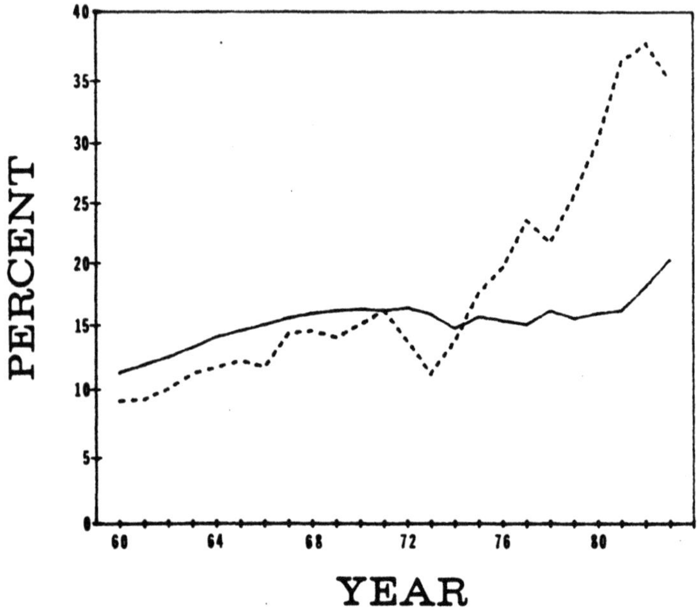

LEGEND ———— Debt/Asset Ratio
-------- Interest Expense/(Net Cash Income + Interest)

From 1960 to 1972 the debt-to-asset ratio rose gradually to a peak of 18.9 percent, after which it stabilized in the 15.5 to 16.5 percent range through about 1979. This stabilization occurred while the value of farm assets and debt rose at unprecedented rates. More recently the debt-to-asset ratio has increased from 16.1 percent in 1979 to 20.7 percent in 1985. The recent sharp rise in the farm sector debt-to-asset ratio suggests a drop in the solvency of farmers as a whole.

The debt-to-asset ratio has also been used as a measure of the debt burden of the sector. It can be misleading, however, as an index of the debt burden. Figure 2.7 illustrates this by showing that the debt-to-asset ratio in the seventies was essentially flat-to-declining, while the ratio of interest to net cash income rose sharply. The rapid rise in farm asset values during this time effectively masked the surge in the debt burden that occurred during that period.

Summary

Analysis of the aggregate financial data suggests a deterioration in the financial well-being of the agricultural sector over the past ten to fifteen years. Farm income has become more volatile and most measures of real farm income show a decline during the period.

The sector debt burden has risen dramatically since the early seventies, sparked initially by low real interest rates, high rates of inflation, high incomes and expectations of high incomes in the future. High interest rates in the eighties and low income have led to debt repayment problems. These factors, coupled with declining asset values, kept the debt burden high in spite of a slow-down in debt growth.

The decade of the eighties finds the farm sector in the throes of a financial adjustment which has been precipitated by unfulfilled expectations of rapidly growing export demand for agricultural commodities and shifts in domestic monetary and fiscal policies that have generated high real interest rates. The result has been a downward adjustment in farm income expectations leading to lower farmland values, declining equity, depressed rates of return, and lower levels of investment.

The conditions described above pose a very difficult setting in which to make farm policy. Agricultural resource prices, particularly that of farmland, appear to be adjusting downward in order to bring the sector back to a more profitable footing. However, the adjustment process has produced considerable stress among certain segments of the sector.

Policies designed to boost commodity prices and incomes are costly and have come under close scrutiny in terms of their effectiveness and contributions to the huge budget deficits that have plagued the economy in the early to mid-eighties. Farm policies must be well focused in the future in order to be effective and cost efficient. However, a detailed understanding of the financial structure of the sector is necessary to focus such policies. In an effort to further that understanding, the next chapter will provide a more disaggregated view of the financial structure of agriculture.

REFERENCES

U.S. Department of Agriculture, Economic Research Service, of Economic Indicators of the Farm Sector: Income and Balance Sheet Statistics, 1983, September 1984.

Miller, Thomas A., Jerry A. Sharples, Robert M. House, and Charles V. Moore, "Increasing World Grain Market Fluctuations: Implications for U.S. Agriculture," USDA, ERS, Agriculture Economic Report No. 541, October, 1985.

3

A Financial Analysis of the Farm Sector: Distributional Characteristics and Nonfarm Comparisons

Stephen C. Gabriel

The previous chapter described the farm sector as experiencing difficulties during the 1980s in terms of several key financial variables. However, agriculture in the United States is characterized by diversity--geographic and climactic, product, and organizational. This chapter will take a close look at how various income and balance sheet items are distributed by sales class, type of farm, and region. The farm sector will also be compared with financial performance indicators of nonfarm business sectors.

Farm and Off-Farm Income by Sales Class

The economic activity of the farm sector during the early 1980s has been dominated by about 300,000 farms with sales of $100,000 or more--12.1 percent of all farms in th U.S. This fact is best illustrated by considering the distribution of various income components by sales class (Table 3.1). In 1983, for example, 66.8 percent of total farm sector cash receipts were generated by farmers with $100,000 or more in sales. One percent of all farmers, those with over $500,000 in sales, generated almost 30 percent of total cash receipts. On the other end of the spectrum, 60 percent of the farms, those with sales less than $20,000, receive only seven percent of total cash receipts.

Large commercial farms do not account for as large a percentage of total production expenses as they do cash receipts. In 1983, farms with sales over $100,000 accounted for almost 59 percent of total production expenses. The efficiencies of large scale farming explain this fact with the largest farms (sales of

TABLE 3.1: DISTRIBUTION OF NUMBER OF FARMS AND SELECTED INCOME COMPONENTS BY VALUE OF SALES CLASS, 1982

	$500,000 and over	$200,000–$499,999	$100,000–$199,999	$40,000–$99,999	$20,000–$39,999	$10,000–$19,999	$5,000–$9,999	$2,500–$4,999	Under $2,500
				——Percent——					
Number of Farms	1.0	3.5	7.5	16.1	11.4	11.8	13.7	13.6	21.4
Cash Receipts	28.9	18.6	19.3	20.0	6.4	3.2	2.0	1.0	0.6
Production Expenses	22.3	17.5	18.9	20.8	7.5	4.5	3.3	2.4	2.8
Net Farm Income	48.4	18.5	17.0	14.6	2.9	-0.3	-0.2	-0.8	-0.7
Off-Farm Income	1.6	2.9	5.1	10.5	9.0	12.3	15.9	15.9	26.8

Source: U.S. Deptartment of Agriculture; 1984

$500,000 or over) requiring an average of only 70 cents in production expenses to generate a dollar of cash receipts. This compares to higher values for other sales classes: 85 cents (sales of $200,000-$499,999), 89 cents (sales of $100,000-$199,999), 94 cents (sales of $40,000-$99,999) and more than a dollar for all remaining sales classes.

As one might expect, the distribution of net farm income, before inventory adjustment, is heavily skewed in favor of the higher sales classes. In 1983, almost 50 percent of net farm income was earned by farms with sales in excess of $500,000. The average net farm income for these farms was $567,000. About 85 percent of sector net farm income was earned by farms with sales of $100,000 or more, averaging about $82,000 per farm. These per farm figures are averages and within each sales class the distribution of incomes may vary widely, ranging from high to negative.

Farms with sales of less than $20,000 had low to negative net farm incomes as a whole in 1983. These same farms, however, earned 71 percent of total off-farm income for the sector. Obviously these farmers, as a group, do not rely on farming for their livelihood. Yet they represent 60 percent of what has been defined as "farmers".

One sales class stands out as possibly experiencing unique difficulties. Farmers in the $20,000 to $39,999 sales class had the lowest average income from farm and off-farm sources in 1983--$16,493. The next lowest income of $18,324 occurred in the $10,000 to $19,999 sales class. These operations may be large enough that the farmer is making a go of it at full-time farming. Or, such farms may be too far away from towns so that off-farm employment is unavailable and the farmer does not have the wealth to expand operations to a financially viable size. Also, these farms may be located primarily in areas which make expansion difficult or impossible due to soil characteristics or other topographical and economic factors. The problems faced by these farmers may well be best addressed through rural development policies rather than farm credit and other agricultural policies.

The Farm Sector Balance Sheet by Sales Class

The farm sector balance sheet is also dominated by farms in the largest sales class, but not to the extent of farm income. Farmers with sales of $100,000 or more own over 43 percent of total farm sector assets (Table 3.2). The very small farms (sales of less than $20,000), however, own 22.3 percent of all farm assets, a substantially greater percentage than their 6.8 percent share of cash receipts.

The largest farms owe about 57 percent of total farm debt while the very small farms owe only 13.3 percent. These distributions are reflected in the debt-to-asset ratio of each sales class. The average debt-to-asset ratio increases as the sales class increases, ranging from almost 12 percent for farms with sales under $10,000 to 38.1 percent for those in the highest sales class. The sector average debt-to-asset ratio is 20.8 percent in Table 3.2.

Farm equity is distributed about the same as assets, with the largest farms holding about 40 percent of the wealth of the sector. Hence, although small farmers contribute relatively little to total farm income as a group, they hold an important share of total assets and equity--20 to 25 percent.

Regional Income and Balance Sheet Comparisons

In this section, the farm sector is divided into ten regions for the purpose of discussing how various parts of the country participate in agricultural production. Each region's share of total U.S. farm assets and debt as well as their contribution to total U.S. farm income is described.

Regional Balance Sheet Comparisons

The Corn Belt is the dominant region of the country in terms of its contribution to the balance sheet of agriculture. In January 1984 it accounted for about 23 percent of total farm assets (Table 3.3). Its share of total farm debt and equity was about the same. The sharp slide in farmland values in the Corn Belt which began in 1980 caused about a five percentage point drop in this region's share in the value of total farm assets.

TABLE 3.2: DISTRIBUTION OF SELECTED COMPONENTS OF THE FARM SECTOR BALANCE SHEET BY VALUE OF SALES CLASS, JANUARY 1, 1984

Items	$500,000 and over	$200,000-$499,999	$100,000-$199,999	$40,000-$99,999	$20,000-$39,999	$10,000 $19,999	Under $10,000
			-----Percent of Total-----				
Real Estate Assets	9.2	14.6	19.4	24.2	10.4	6.9	15.2
Total Assets	10.0	13.8	19.3	24.2	10.3	6.8	15.5
Total Farm Debt	18.3	18.3	20.1	22.4	7.6	4.7	8.6
Farm Equity	7.8	12.7	19.0	24.8	11.0	7.4	17.3
			--------Percent--------				
Debt-to-Assets	38.1	27.5	21.8	19.1	15.3	14.4	11.6

Source: U.S. Dept. of Agriculture, 1984

TABLE 3.3: REGIONAL DISTRIBUTION OF FARM ASSETS DEBT AND EQUITY, 1980 AND 1984

Region	1980 Assets	1980 Debt	1980 Equity	1984 Assets	1984 Debt	1984 Equity
	Percent					
Northeast[1]	4.7	4.4	4.8	4.8	4.4	4.9
Lake States[2]	9.1	10.3	8.8	8.9	10.9	8.4
Corn Belt[3]	26.9	23.6	27.6	22.3	22.5	22.2
North Plains[4]	12.4	14.0	12.1	11.6	14.5	10.8
Appalachian[5]	6.9	6.8	6.9	7.0	6.2	7.2
Southeast[6]	5.9	6.4	5.8	5.8	6.0	5.7
Delta States[7]	5.2	5.2	5.2	5.4	5.3	5.5
South Plains[8]	10.7	9.1	11.0	13.4	9.2	14.6
Mountain[9]	9.3	9.3	9.3	9.7	9.0	9.9
Pacific[10]	8.8	10.8	8.4	11.1	11.9	10.9
United States	100.0	100.0	100.0	100.0	100.0	100.0

[1] ME, NH, VT, MA, RI, CT, NY, NJ, PA, DE, MD
[2] MI, WI, MN
[3] OH, IN, IL, IA, MO
[4] ND, SD, NB, KS
[5] VA, WV, NC, KY, TN
[6] SC, GA, FL, AL
[7] MS, AR, LA
[8] OK, TX
[9] MT, ID, WY, CO, NM, AZ, UT, NV
[10] WA, OR, CA

The next most important regions in terms of values of assets are the Southern Plains, Northern Plains, and Pacific States with 13.4, 11.6, and 11.1 percent of U.S. total farm assets, respectively. The regions with the smallest shares of total farm assets include the Northeast, the Delta States, and the Southeast with 4.8, 5.4, and 5.8 percent of the total, respectively. The regional distributions of farm debt and equity are similar to that of total assets.

The composition of farm assets also varies by region. Although real estate accounted for about 75 percent of total farm assets nationwide, this percentage varies considerably from region to region. For example, real estate in the Corn Belt makes up about 74 percent of total assets, close to the national average. The percentages in the Northeast, Lake States, and Northern Plains are between 66 and 69 percent. The real estate percentage reaches 82 and 83 percent in the Southern Plains and Pacific States, respectively. Regions where real estate is a high percentage of total value of assets appear be those where irrigation equipment would be more widely used than in other parts of the country.

Debt-to-asset ratios ranged from 27.1 percent in the Northern Plains to 14.9 percent in the Southern Plains on January 1, 1984 (Table 3.4). In 1980 debt-to-asset

TABLE 3.4: REGIONAL DEBT-TO-ASSET RATIOS, 1980 AND 1984

Region	1980	1984	Percent Change
	-------Percent-------		
Northeast	16.0	20.0	25
Lake States	19.4	26.3	36
Corn Belt	15.0	21.9	46
North. Plains	19.3	27.1	40
Appalachian	16.7	19.3	16
Southeast	18.5	22.6	22
Delta States	17.2	21.2	23
South. Plains	14.5	14.9	3
Mountain	17.1	20.0	17
Pacific	21.0	23.2	10

See Table 3.3 for definition of regions
Source: U.S. Department Of Agriculture, 1985

ratios ranged from 21 percent in the Pacific States to 14.5 percent in the Southern Plains. Farmers were more highly leveraged in 1984 in all regions of the country than they were in 1980. However, the increase in leverage varied in degree for different regions of the country. The largest percentage increase in debt-to-asset ratios occurred in the Corn Belt, jumping 46 percent in four years. The debt ratio in the Northern Plains grew 40 percent during this period. The smallest percentage increases occurred in the Southern Plains and the Pacific States where the ratios increased only 3 percent and 10 percent, respectively.

The large increases in the debt-to-asset ratios in the Corn Belt and Northern Plains were precipitated by very weak farmland markets, which pushed the value of total assets down 15 percent in the Corn Belt and four percent in the Northern Plains. In the Southern Plains and the Pacific States, on the other hand, relatively strong farmland markets resulted in healthy increases in total farm asset values--28 percent and 29 percent, respectively. Strength in land markets in the Southern Plains stems from considerable oil revenue, which is generated on many farms and ranches in the region. The Pacific States benefit from a highly diversified agriculture which includes many profitable specialty crops such as fruits, nuts, and grapes.

In each region debt grew rapidly, increasing from 42 percent in the Pacific States to 19 percent in the Appalachian States. Hence, with assets' growth lagging that of farm debt in every region of the country, the debt-to-asset ratios climbed.

Regional Farm Income Comparisons

The distribution of cash receipts among the regions of the country points once again to the Corn Belt as the highest ranking region, followed by the Pacific States and the Northern Plains. The distribution of total sector production expenses is roughly the same as that of cash receipts. Both the 1979 and 1984 distributions are comparable.

Examination of the distributions of net farm income leads to some interesting discoveries. In 1979 and 1983, the Pacific States captured the largest and second largest percentage of total net farm income--17.4 percent and 19.4 percent in the two years, (Table 3.5). This was done with only 8.8 percent and 11.1 percent of total farm assets in those years. The implication is that farmers in the Pacific States earn relatively high rates of return on assets. Corn Belt farmers' share of total net farm income was 16.5 percent and 22.2 percent in 1979 and 1983, respectively.

TABLE 3.5: DISTRIBUTION OF CASH RECEIPTS, PRODUCTION EXPENSES, AND NET FARM INCOME BY REGION, 1979 AND 1983

Region	Cash Receipts		Production Expenses		Net Farm Income	
	1979	1983	1979	1983	1979	1983
	-----------------Percent-----------------					
Northeast	6.0	6.7	6.3	6.9	5.5	4.9
Lake States	9.2	10.4	9.2	10.3	10.1	11.1
Corn Belt	21.1	21.0	22.4	20.9	16.5	22.2
North. Plains	12.8	12.0	13.1	12.9	10.2	9.9
Appalachian	6.9	7.3	7.1	7.5	7.9	7.2
Southeast	7.5	7.8	6.8	6.7	10.0	10.6
Delta States	5.2	5.1	5.0	5.0	6.3	6.6
South. Plains	10.2	8.4	10.1	9.6	10.3	5.9
Mountain	8.1	7.5	8.7	8.2	5.5	4.2
Pacific	12.6	13.2	11.1	11.7	17.4	17.0

See Table 3.3 for definitions of regions.
Source: U.S. Department of Agriculture, 1985

An interesting ratio which is indicative of the efficiency with which revenues are generated is the inverse gross ratio, defined here as cash receipts divided by production expenses. Simply stated, it provides the average cash receipts earned per dollar of production expenses. In spite of its predominance in other categories, the Corn Belt does not rank particularly well when it comes to being "revenue efficient" in 1979 and 1983 (Table 3.6). This, obviously, reflects a relatively high cost production process and indicates that, as a whole, farmers in such regions are particularly vulnerable to unstable commodity prices. The regions which are most "revenue efficient" are the Pacific States led by California and the Southeast, due particularly to the performance of Florida and Alabama.

TABLE 3.6: RATIO OF CASH RECEIPTS TO PRODUCTION EXPENSES BY REGION, 1979 AND 1983

Region	1979	1983
	-------Percent-------	
Northeast	1.06	1.01
Lake States	1.12	1.04
Corn Belt	1.05	1.03
North. Plains	1.09	0.96
Appalachian	1.07	1.00
Southeast	1.24	1.20
Delta States	1.18	1.05
South. Plains	1.12	0.90
Mountain	1.05	0.94
Pacific	1.27	1.16
United States	1.12	1.03

See Table 3.3 for definitions of regions.
Source: U.S. Department of Agriculture, 1985

These regions seem to benefit from profitable specialty crops, such as fruit, vegetables, and nuts.

Farm Income by Type of Farm

The agricultural sector is far from a homogeneous collection of farms as many people might think. Indeed, the farm sector is diverse, comprised of many different types of farms with different income characteristics. This diversity is shown in part by the income position of different types of farms.

The distribution of net farm income, before inventory adjustment, among various types of farms is provided for 1978 and 1983, showing the relative importance of the major categories of farms in generating farm income in those years (Table 3.7). Cash grain farms (corn, wheat, soybeans, rice, and other cash grain) contributed 34 percent of the total net farm income generated by crops

TABLE 3.7: DISTRIBUTION OF NET FARM INCOME BEFORE INVENTORY ADJUSTMENT BY FARM TYPE, 1978 AND 1983

Farm Type	Year	
	1978	1983
	--Percent of Crop Subsector--	
Cash Grain	33	33
Cotton	6	10
Tobacco	7	4
Other Field Crops	8	17
Vegetable and Melon	14	14
Fruit and Tree Nut	17	6
Horticulture Specialties	9	10
	--Percent of Livestock-- --Subsector--	
Cattle, Hogs, and Sheep	47	35
Dairy	34	46
Poultry and Eggs	13	23
	---Percent of Total---	
All Crops	52	67
All Livestock	48	33

Source: Somwaru

in 1978. Their share in 1983 was 33 percent. Next in importance were vegetable and melon farms, contributing 14 percent to the net farm income generated by crop farms. In January 1978, fruit and tree nut farms and farms producing other field crops (sugar, irish potatoes, and other) had 17 percent and eight percent shares of total crop sector income. By 1983, those shares had reversed. Farms producing cotton and tobacco earned between four and ten percent of total crop net farm income. Although these types of farms contribute a relatively small share of total net farm income from crops, they are of great importance regionally (e.g., cotton in the Southern Plains and parts of California and tobacco in the Southeast). Also, if one were to compare the income earned by tobacco and cotton farms with that of other farm categories on a per commodity basis, they would compare quite favorably.

Livestock farms include cattle, hogs and sheep farms, dairy farms, and poultry farms. In 1978 the shares of total net farm income generated by livestock farms were 47, 35, and 14 percent, for cattle, hogs, and sheep farms; dairy farms; and poultry farms, respectively. The relative importance of cattle, hogs, and sheep farms and dairy farms reversed in 1983 as the net farm income of the former dropped 52 percent, while that of dairy farms dropped only 3 percent. In 1982 dairy farms contributed over 60 percent of total net farm income in the livestock sector.

The Farm Sector Balance Sheet by Farm Type

The nation's farm assets are concentrated primarily among cash grain farms and cattle, hog, and sheep farms. These types of farms employed about 66 percent of total farm sector assets on January 1, 1980 (Tables 3.8 and 3.9). Among the crop farms three types of operations stand out in terms of their share of total crop subsector assets: cash grain, fruit and tree nut, and other field crops, with 58, 12, and 10 percent of farm assets owned by crop farmers. This ranking is similar to that of farm income as discussed earlier. In the livestock subsector, cattle, hogs, and sheep farms owned 73 percent of farm assets followed by dairy farms and poultry and egg farms with 18 percent and 4 percent, respectively (Table 3.9). The distributions of the individual asset items are roughly similar to those of total assets.

The distribution of farm debt suggests differing financial structures among the various types of farms. Debt-to-asset ratios ranged from 14 percent for fruit and tree nut farms and tobacco farms to 24 percent for cotton farms. While it is unlikely that the debt-to-asset ratios which existed in 1980 are indicative of today's conditions, they do indicate the diversity which exists in the sector at a given point in time.

TABLE 3.8: DISTRIBUTION OF SELECTED COMPONENTS OF THE FARM SECTOR BALANCE SHEET BY TYPE OF CROP FARM, JANUARY 1, 1980

Item:	Cash Grain	Cotton	Tobacco	Other Field Crops	Vegetable and Melon	Fruit and Tree Nut	Horticulture Speciality	All Crops
Assets:	Percent of Crop Subsector							Percent of total
Land	59	4	4	9	2	14	2	49
All Buildings	50	2	8	12	4	12	6	40
Machinery and Equipment	65	5	5	9	3	6	2	56
Crops	81	3	1	6	1	3	1	64
Livestock and Poultry	60	3	6	13	1	1	0	14
Financial Assets	47	6	5	11	4	16	4	44
Total Assets	58	4	4	10	3	12	3	46
Claims:								
Total Farm Debt	61	5	3	9	3	9	2	49
Equity	58	4	5	10	3	12	3	45
	Percent							
Debt-to-Assets	19	24	14	16	20	14	16	18

Source: U.S. Department of Commerce:

TABLE 3.9: DISTRIBUTION OF SELECTED COMPONENTS
OF THE FARM SECTOR BALANCE SHEET
BY TYPE OF LIVESTOCK FARM, JANUARY 1, 1980

Item	Cattle, Hogs, Sheep	Dairy	Poultry and Eggs	All Livestock
		Percent of Livestock Subsector		Percent of Total
Assets:				
Land	79	14	3	51
All Buildings	64	21	8	60
Machinery and Equipment	66	26	4	44
Crops	69	24	2	36
Livestock and Poultry	65	27	2	86
Financial Assets	78	12	4	56
Total Assets	73	18	4	54
Claims:				
Total Farm Debt	70	20	5	51
Equity	74	18	3	55
	-------------Percent-------------			
Debt-to-Assets	15	18	22	16

Source: U.S. Department of Commerce:

Income statements and balance sheets by farm type corroborate the regional data which point out the lower "revenue efficiency" of the Corn Belt. Cash grain farmers claimed only about 30 percent of crop subsector income, although they owned 58 percent of that group's assets in 1980. Cattle, hog, and sheep farmers likewise earned a proportionally lower share of livestock sector income. These types of farms are dominant in the Corn Belt, which suggests that capital gains and losses have relatively greater effect on the overall financial performance of these farms.

Farm and Nonfarm Businesses: A Comparative Financial Analysis

The financial performance of the farm sector compared to nonfarm businesses can be evaluated using the flow of funds accounts of the Board of Governors of the Federal Reserve System. Income and balance sheet trends for the two sectors, farm business and nonfarm noncorporate business, will be compared for the purpose of assessing the financial performance of agriculture since 1960 relative to another sector of the economy. Nonfarm noncorporate businesses are selected for comparison with farm businesses because they are the most comparable sector in the flow of funds accounts in terms of organizational structure and access to outside equity capital.

Net Income

Trends in farm income do not compare favorably with those of nonfarm noncorporate businesses since 1960 (Figure 3.1). The overall trend in real net farm income has been negative from 1960 to 1983. The slope of the trend line is negative in each decade represented during this period. Real net income in the nonfarm noncorporate business sector, however, trends upward from 1960 to 1983. If this period is segmented into the sixties and the seventies, both show a positive trend for nonfarm noncorporate businesses. Real net income in the nonfarm sector has been declining, however, since 1979.

A five year moving coefficient of variation of real net income indicates greater income variability in recent years for both farm and nonfarm businesses (Table 3.10). However, the variability of real farm income has been consistently higher than that of nonfarm noncorporate businesses since 1960. The data suggest that risk has increased more for farmers as a whole during the last ten to fifteen years.

Indices of the Debt Burden

As discussed in the previous chapter, the farm sector debt-to-asset ratio rose gradually through the early to mid-sixties, then stabilized during the late sixties to early seventies, settling before rising sharply in the eighties. The debt-to-asset ratio

FIGURE 3.1: REAL NET INCOME

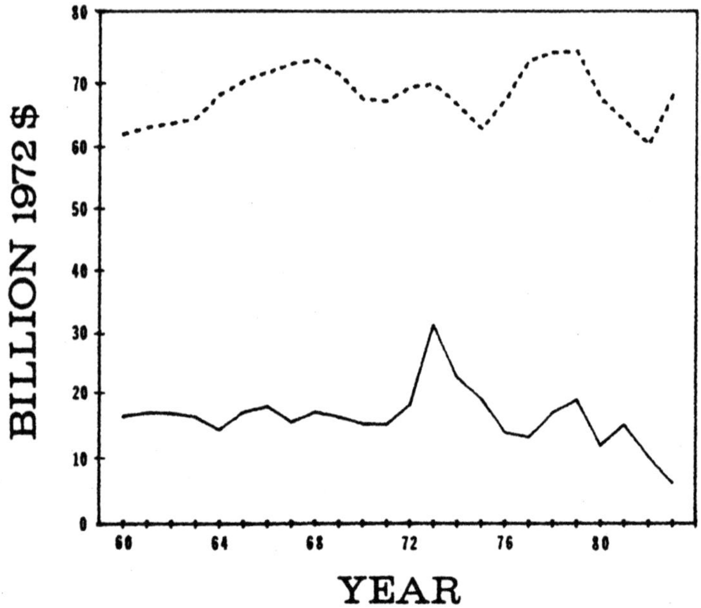

LEGEND ——— Farm Business Real Net Income

.... Nonfarm Noncorporate Real Net Income

exhibited stability during the seventies in spite of rapid growth in farm debt due to a boom in farm real estate values. The sharp rise in the debt-to-asset ratio in the early eighties was precipitated by a substantial drop in farmland values.

Nonfarm businesses experienced a rapid rise in their

debt-to-asset ratio from 1960 to 1973 as total liabilities grew at a rate of 13 percent per year, while assets grew at only a five percent rate (Figure 3.2). The

TABLE 3.10: VARIABILITY OF REAL NET INCOME OF FARM AND NONFARM NONCORPORATE BUSINESSES, 1964-83

Year	Farm Businesses	Nonfarm Noncorporate Businesses
	-------Percent-------	
1964	5.8	3.4
1965	6.1	4.4
1966	7.0	4.8
1967	7.4	4.4
1968	7.6	2.7
1969	4.8	1.6
1970	6.0	3.1
1971	4.2	3.9
1972	7.0	3.8
1973	29.9	2.3
1974	27.8	2.0
1975	24.4	3.8
1976	26.0	3.8
1977	31.3	5.3
1978	19.5	6.6
1979	14.8	6.9
1980	17.3	4.8
1981	16.7	6.3
1982	21.9	8.7
1983	34.7	7.4

Source: Five moving year coefficient of variation of real net income. Board of Governors of the Federal Reserve System: Flow of Funds Account, Third Quarter 1983 and Previous issues.

debt-to-asset ratio stabilized at a plateau of 29 to 30 percent before dropping sharply to about 23 percent in 1982. Several interesting observations can be made regarding this comparison. Nonfarm businesses apparently have the ability to make rather substantial adjustments in their capital structure over time. The farm sector has demonstrated an ability to make only gradual adjustments. Two reasons for this difference in capital structure flexibility come to mind. The farm sector has a particularly illiquid asset structure, consisting

FIGURE 3.2: DEBT/ASSET RATIO

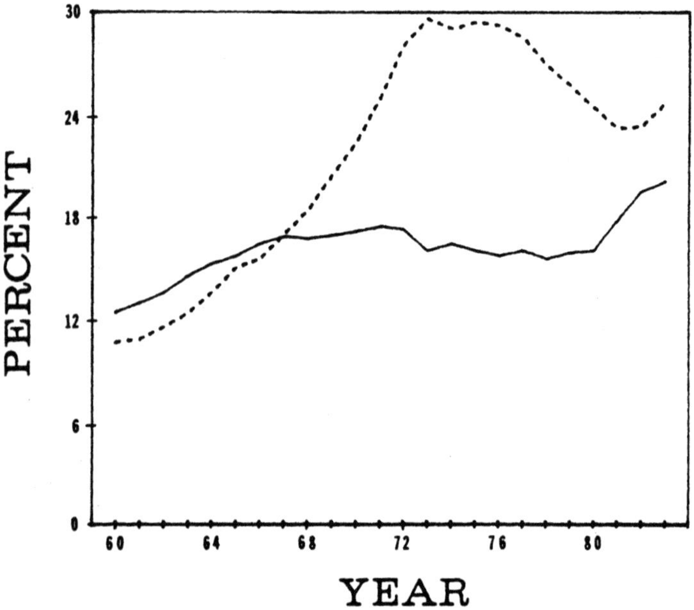

LEGEND ———— Farm Business Debt-to-Asset Ratio
.... Nonfarm Noncorporate Debt-to-Asset Ratio

primarily of farmland. Also, equity markets in agriculture are grossly underdeveloped making it difficult for farmers to obtain outside equity. This is due in part to hostile attitudes on the part of farm groups toward the prospects of "corporate" ownership of farms. Some states have laws which prohibit such activities. While nonfarm noncorporate businesses, by definition, have no access to formal equity markets, it is likely that they utilize informal equity markets to a greater extent due to their proximity to urban areas, affording them greater access to available capital. Finally, nonfarm businesses have experienced periods of sustained real growth in net income, making internally generated funds available for their financing needs. Such a period occurred from 1975 to 1979, which allowed for a considerable decline in debt-to-asset ratios. Since 1960, the farm sector has not enjoyed such a luxury.

The ratio of debt to net income provides a quite different perspective on the debt burden in the farm and nonfarm sectors (Figure 3.3). With the exception of 1973, a record income year for farmers, farm debt per dollar of farm income has been consistently higher than the comparable ratio for nonfarm noncorporate businesses --and the difference has been growing since 1973. The performance of the debt-to-net-income ratio in agriculture contrasts sharply with that of nonfarmers during the last ten years or so. In the farm sector, the ratio rose sharply at an annual rate of 14 percent on trend, a much higher rate of growth than in the sixties. This compares with less than a four percent growth rate in the nonfarm sector. Interestingly, while the farm sector debt-to-asset ratio showed relative stability over the last twenty years or so compared to a more volatile debt-to-asset ratio in the nonfarm sector, the reverse is true for the debt-to-net-income ratio.

The previous discussion provides some perspective to intersectorial comparisons of debt-to-asset ratios. The risk associated with any given debt-to-asset ratio must be evaluated in terms of the stability and level of the income stream with which the debt will be serviced. Two firms with the same debt-to-asset ratio could be carrying much different debt burdens if their income streams differ in level and/or volatility. That is, the risk of default could be much higher for one than the other.

It is not uncommon to see the farm sector debt-to-asset ratio compared to that of nonfarm businesses. Usually it is suggested that since the farm sector ratio is lower than that of nonfarm businesses, the sector is not highly leveraged as a whole. However, considering the relatively high volatility of farm income and the large amount of debt per dollar of net income, the farm sector ought to maintain a lower debt-to-asset ratio compared to nonfarm businesses. In other words, nonfarm businesses could well have a higher debt-to-asset ratio than farmers, and still be exposed to the same or less total risk of default because of the difference in economic and financial characteristics between the sectors.

Net Worth

The real net worth (equity) of both farm and nonfarm noncorporate businesses followed a roughly flat trend from 1960 to 1971 (Figure 3.4). By the middle seventies

FIGURE 3.3: DEBT/NET INCOME RATIO

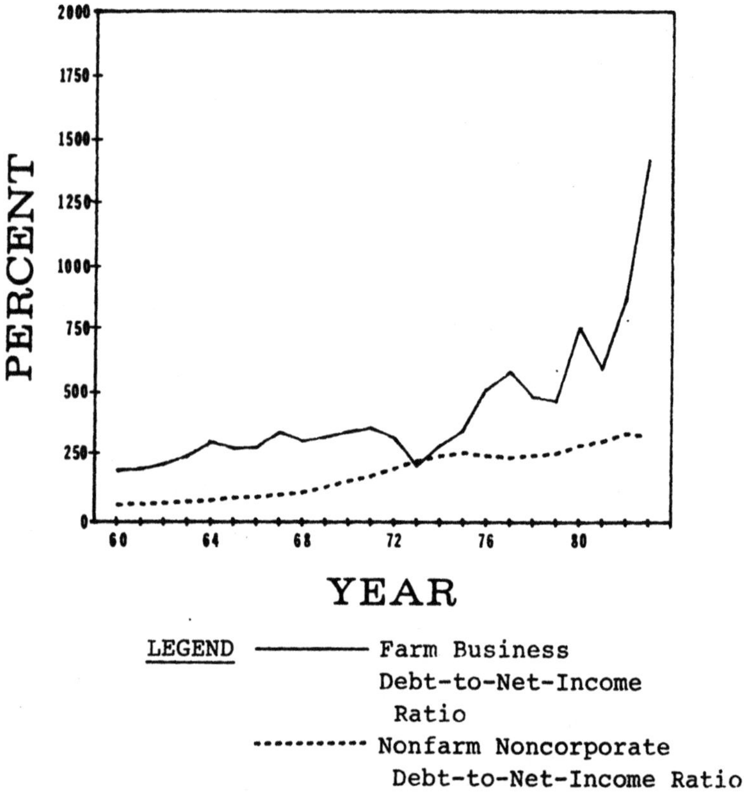

real equity was growing at a rapid pace in both sectors, resulting in a substantial accumulation of wealth during this period. By 1980 the boom in farmland values had expired and was followed by several years of decline. This caused severe financial difficulties for those farmers who had employed substantial financial leverage to make large capital purchases in the seventies.

Nonfarm noncorporate real equity has not begun to fall off, although its rate of growth has moderated considerably.

FIGURE 3.4: REAL NET WORTH

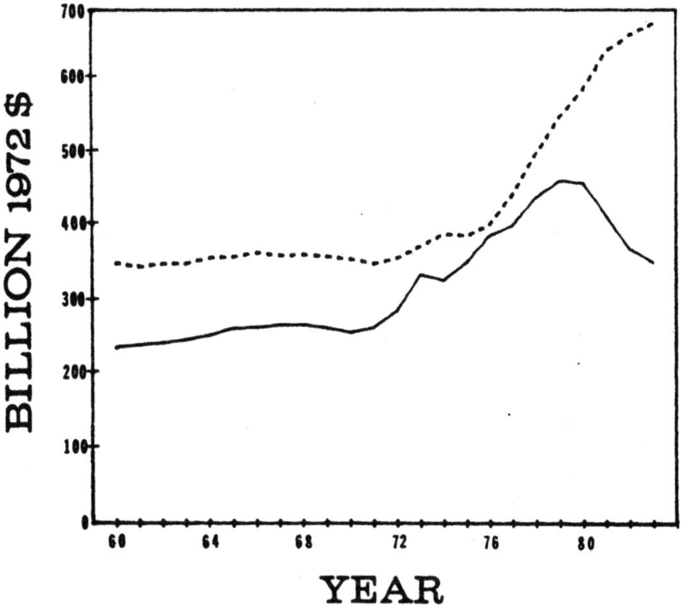

LEGEND ——————— Farm Business Net Worth
··········· Nonfarm Noncorporate Net Worth

Summary

It appears that, as a group, nonfarm noncorporate businesses are stronger financially than their farm counterparts. Farm income is more variable, requiring more effective risk management. The farm sector also carries a high debt burden with about 2.7 times as much debt per dollar of net income.

The diversity of agriculture makes the formation of farm policy, including farm credit policy, a complex matter. From year to year, the well-being of farmers producing a given commodity depends not only on weather and general economic conditions, but on market developments for other commodities, as well as conditions in other countries. Trade developments in another country may also affect different commodities differently, depending on trade patterns. Hence, record prices for feed grains such as corn and soybeans may result in high incomes for cash grain farms, but they translate into record high feed costs for the livestock sectors.

Farm credit policies that are linked to weather such as emergency disaster loans, may direct a disproportionate share of benefits toward farmers whose greatest source of risk is weather related, while farmers whose main concern is price risk derive relatively little benefit.

REFERENCES

Board of Governors of the Federal Reserve System, Flow of Funds Account, Third Quarter, 1983 and previous issues.

Somwaru, Agapi, "An Analysis of Economic Trends in Farm Income and Structure by Type of Farm, 1959-82," U.S. Department of Agriculture, Economic Research Service, forthcoming.

U.S. Department of Agriculture, Economic Research Service, Economic Indicators of the Farm Sector: Income and Balance Sheet Statistics, 1983, September, 1984.

U.S. Department of Agriculture, Economic Research Service, Economic Indicators of the Farming Sector: State Income and Balance Sheet Statistics, 1983, January, 1985.

U.S. Department of Commerce, Bureau of the Census, 1979 Farm Finance Survey, 1978 Census of Agriculture, July, 1982.

4

The Capital Markets for Farm Firms

Dean W. Hughes

The purpose of this chapter is to describe ways farmers can obtain financing. There are really only two sources of funds for buying any investment: equity--using your own money or, debt--using someone else's money. Leasing provides a method of controlling an asset without ownership. However, many long term lease agreements with fixed payments affect the financial condition of the lessee in the same way as a long-term loan.

Whenever the returns on an investment exceed the interest rate being charged for the use of others' money, some use of debt makes sense. There are, however, more things to consider than simply comparing rates of returns on assets and interest rates on debt. Debt must be repaid, and such payments become a fixed expense, required regardless of the actual returns on the investment in a given year. With variable interest rate loans, additional variation in returns to equity can be introduced if the returns to the investment do not increase when interest rates rise. Thus, while a higher proportion of debt financing on a profitable project will increase the returns on the money invested, it will also increase the variability, or risk, of those returns.

Given the relationship between higher returns and added risk involved in leveraging, there is usually an optimal division between debt and equity financing for businesses. With less than optimal debt, firms give up more return than the benefit they gain by the reduction in risk. With more debt than the optimal, firms gain too much risk for the available additions to returns.

In many instances, farmers are unable to achieve an optimal ratio of debt to equity since they have constraints on their ability to generate equity capital.

Young farmers are perhaps the best example of this problem. A large amount of capital is required to start farming. Without inheriting a farm, young farmers usually end up with large amounts of debt and consequently face high risks. Increases in equity can come either through retained earnings--reinvesting the profits of past investments, or by selling stock--selling part of the ownership of the business. As of 1978, over 99 percent of all farms were still owned by one individual or one family. Thus, most farms must still rely solely on reinvesting profits to accumulate the equity needed to move them toward an optimal amount of debt.

Since equity is an important constraint in farm financial decisions, the sources of farm equity are discussed first. Private sources of farm debt will then be described, followed by government lenders to agriculture. The chapter will end with a discussion of leasing as an alternative to debt financing.

Sources of Farm Equity

Farm sector equity comes from two sources: total returns from farming that are not withdrawn by owners, and funds from outside the sector. By far the largest increases in farm equity have come from retaining profits and capital gains. Individual farmers can also gain equity when farm assets are transferred to them by way of gifts and inheritances or by selling equity shares to other farmers, but this does not increase the equity of the sector as a whole. Small quantities of equity funds flow into the sector through a variety of channels including merging farm operations with other agribusinesses, individual investors buying farm assets, syndicated partnerships and even some funds coming from the sale of corporate farm stock.

Retained Earnings

The total return to farming is comprised of current profits and capital gains on owned assets. Current profits have historically been low in farming. However, capital gains on farm assets have, in many years, made total returns to farm businesses meet or exceed the returns in many other sectors.

Measuring retained earnings out of current income

for the farm sector has been and continues to be a very difficult, if not impossible, exercise. Basic information such as farmers' withdrawals for family living is not collected with sufficient frequency by any central agency to generate reasonable data. One would expect that periods of unanticipated high profits, such as 1973-74, provided substantial increases in farmers' equity through the retention of current income. In other times, such as 1980-85, when profits were historically low, farm equity declined, as withdrawals likely exceeded earnings.

Capital gains on farm assets can be measured properly, but not simply. After adjusting for inflation, farm assets grew from $251 billion in 1950 to $611 billion in 1980. Part of this increase was due to increases in the numbers of tractors, trucks, and other farm equipment and, therefore, not due to capital gains. Even in the category of farm real estate, improvements in structures, fences, etc., have led to value increases which were not capital gains.

Farm assets grew from $251 billion in 1950 to $582 billion in 1980, after adjusting for inflation and investments in assets. Part of these gains have accrued to farmers, and the rest to nonoperator landlords who own farm assets but are not directly involved with farm operations. Periodic surveys have indicated that about two-thirds of farm assets were owned by farm operators. Applying this percentage to the real capital gains of the sector suggests that farmer operators' real wealth grew by $221 billion between 1950 and 1980. Nonoperator landlords' real wealth increased by $110 billion over the same period.

Unrealized capital gains can not be used to pay expenses or buy consumption goods. Of course, cash can be generated by using the capital gains as collateral for additional borrowing. Unless conditions change, however, such actions simply postpone the time when there is insufficient cash to meet current commitments. Delay is all that is needed for some discontinuing farmers. Once land is sold and gains are realized, debt can be repaid, leaving the ex-farmer with substantial wealth. For the continuing farmer, however, low current cash returns cannot be completely offset by capital gains without forcing farmers to live in poverty or constantly increasing their debt.

There are also risks of price declines when the capital gains are still unrealized. According to Melichar (1984a), real capital losses in farm real estate amounted to about $150 billion between 1980 and 1983. This loss represented 11.4 percent of 1980 farm assets and 13.6 percent of 1980 equity. When combined with lower farm incomes, such losses have precipitated a crisis in farm lending. Loan losses at agricultural banks have grown rapidly along with rural bank failures. Past reliance on unrealized capital gains in making loan decisions by both farmers and farm lenders has, in hindsight, proven to have been inappropriate.

Other Sources of Farm Equity

Farmers can use numerous other methods to obtain funds from outside the farm sector. Some farms are owned by large corporations and have indirect access to national equity markets. Other farms are linked to agribusinesses through forward contracting of farm output, which can require contributions to working capital. Considerable farm real estate is rented through crop or livestock share agreements. In these rentals the farm operator receives use of the land for a fraction of the eventual production. It is common in such arrangements for the landlord to share some of the input expenses and thus provide funding for an equity participation in the value of output. Dollars can also flow into the sector through partnerships, where some of the partners are not already involved in farming. Finally, some very large farms and ranches are publicly held corporations and can directly access equity funds through the sale of stock in regional and even national exchanges.

Even though there are many avenues for farm equity generation, dollar amounts of equity generated in this way are relatively small. There are several explanations for the low levels of use of external equity by the sector. The costs of incorporation and stock sales including lawyers' fees, improved record keeping and SEC regulation, may be prohibitive for the average size farm. Information about the risks and management arrangements of individual farms may be expensive to obtain and difficult to update. If so, the risk premiums investors would require might make equity financing too expensive. Finally, uninformed inertia in farmers' attitudes may be a factor. They may feel that additional owners would interfere with their management decisions, or they may not feel that they can or should invest the time required to understand alternative means of equity generation.

Private Sources of Farm Debt

The USDA keeps records on four groups of private lenders to farmers. The Farm Credit System is the largest of these groups. The System held 43 percent of farm real estate debt and 19 percent of farm nonreal estate debt at the end of 1984. Individuals and others are second, holding 29 percent of the real estate debt and 18 percent of nonreal estate debt. Banks hold more nonreal estate debt than any other lender, 38 percent at the end of 1984, but have not been very active in real estate lending, holding only 8 percent. Life insurance companies are the fourth group and have concentrated in farm real estate lending, owning 11 percent of such debt at the end of 1984.

Farm Credit System

The Farm Credit System (FCS) is a confederation of federally chartered, borrower-owned financial cooperatives. FCS raises funds in the national money markets and makes loans to farmers, rural banks, and other farmer cooperatives. It is composed of three lending systems, a regulating organization, a government affairs organization and a funding agency. There are twelve district Federal Land Banks (FLBs) and each is owned by local cooperatives called Federal Land Bank Associations (FLBAs). FLBAs are in turn owned by member borrowers who are farmers and ranchers. There are also twelve Federal Intermediate Credit Banks (FICBs), owned principally by local cooperatives called Production Credit Associations (PCAs). PCAs are also owned by farmers and ranchers. Finally, there are twelve district level Banks for Cooperatives (BCs) and one Central Bank for Cooperatives. All the BC's are owned by their member-borrowers. Since they lend to farmer cooperatives, ultimate ownership is by farmers and ranchers. Exceptions occur when rural banks sell farm loans to the FCS, or when loans are made to individuals who own farm assets but are not farm operators. Thus these groups end up owning a small fraction of the system.

Just as commercial banks are regulated by governmental and quasi-governmental agencies (such as the Federal Reserve System), the FCS is regulated by the Farm Credit Administration (FCA). The FCA assures that legal restrictions on the activities of the FCS are followed and acts as a central information source for the system. All of the expenses of the FCA are paid for by the system so no taxpayer expense is involved. Recently the FCS established its own government affairs organization called the Farm Credit Council (FCC). The FCC serves the purpose of representing the System in legislative discussions just as the American Bankers Association and the Independent Bankers Association of America serve the banking industry.

The fiscal agent of the system, the Federal Farm Credit Banks Funding Corporation, is located in New York City and sells the system's securities to investors. Each bank prepares an estimate of its funding requirements for the coming month. The fiscal agent determines the total quantity of funds needed and suggests maturities that are likely to meet the greatest demand and thus require the lowest interest rates. Each bank decides what fraction of their requirements they want funded at the different maturities. The funding corporation packages the securities to meet the requirements of the system and the demands of the market, sells the system's consolidated securities bonds and distributes the funds to the banks.

The FLBs and FLBAs are the largest lenders of farm real estate debt. They were established in 1916 to overcome difficulties in farm real estate transactions. Prior to that time the normal maturity for a real estate loan was about five years and it was difficult for the land to generate sufficient returns to repay both principal and interest in such a short period of time. By bringing to the markets a source of funding that had a longer maturity, the FLBs and the FLBAs were able to meet a previously unserviced need.

The FICBs and PCAs are the second largest lenders of nonreal estate farm debt. The FICBs were established in 1923 to overcome periodic shortages of short and intermediate term debt funds in rural financial markets. They were originally established with the idea that commercial banks would make farm loans, sell the loans to the FICBs, and thus have more funds to lend to other farmers. Bank discounting of farm loans was not widely accepted, so the PCA system was established in 1933 to provide a direct link between national money markets and farm operating loan demand.

The BC system was established in 1933 to provide credit to farmer cooperatives. Farming has been perceived to be a perfectly competitive sector of the economy sandwiched between input suppliers and output processors who all have some degree of influence over prices paid to or received by farmers. Farmer cooperatives allow farmers to join together and exert offsetting market power to preserve the returns to farming against the concentrations of market power in farm input and output markets. Farmer cooperatives have, therefore, been seen as serving a public benefit and have been encouraged by the government. The BCs have contributed to this encouragement by financing the establishment, growth and operations of the various types of farm cooperatives.

Even though the FCS was originally established early in this century by the federal government, it currently operates almost entirely as a private enterprise. All government seed money was repaid by 1968 so no direct government funding is involved in FCS loans. The system still enjoys some advantages of its origins, however. It is classified as a government agency, and thus is able to borrow at interest rates only slightly greater than the U.S. Treasury. The interest on its bonds is generally considered to be free from state and local taxes. And, the Federal Land Banks, the Federal Land Bank Associations and the Federal Intermediate Credit Banks do not pay federal income taxes.

As part of the system's cooperative nature, interest rates on its loans are determined by each bank's average cost of funds. Their operating costs, which include provisions for loan losses, are also included in lending rates. Since most loans have short maturities or carry variable rates, all borrowers generally share the impacts of rising or falling interest rates. Other lenders, particularly large commercial banks, usually charge a borrower the cost of raising the additional funds needed to finance the new loan. Thus only new loans are affected by changes in interest rates in these other institutions.

Average cost pricing has been a mixed blessing to FCS borrowers. The process moderates fluctuations in rates, which is good if interest rates are rising, but a problem in times of declining rates. It also makes borrowers' actions depend on what other existing and potential borrowers are doing. In periods of rising interest rates, for example, new borrowers are likely to join the system since increases in FCS interest rates will lag behind those of other lenders. The funds needed to service the new borrowers must be raised at the new, higher market rates and the average cost of funds for the system will grow more quickly than it would have if there had been no new borrowers. In some sense then, existing borrowers end up paying part of the interest expenses of new borrowers. This leads to a desire on the part of existing borrowers to exclude new borrowers during periods of rising rates. To offset this subsidy, there have been times when the banks have instituted loan origination fees that effectively impose higher rates on new borrowers. It can also be seen that the winning strategy for a farm borrower is to join the FCS when interest rates are going up and leave the system for other lenders when interest rates are declining. This strategy is one that is effective for large borrowers that have several potential sources of funds such as large farmer cooperatives. This "free rider" use of the system increases direct costs to existing borrowers and indirect costs to all system borrowers by increasing the difficulties of predicting loan demands for the banks' management.

Individuals and Others

This is a category in the USDA's accounting system that is difficult to describe. Input suppliers, families and other individuals comprise the majority of the lenders of Individuals and Others (IO) nonreal estate debt. Sellers of farm real estate are the principal lenders of IO real estate debt.

Input suppliers use credit to sell their goods. Since credit activities are not their primary function, the availability of credit from these institutions has been highly variable. Moreover, the cost of using trade credit can vary widely. Some firms provide merchandise and do not expect any return for waiting 30 days or more for payment. Others charge high rates, sometimes disguised in the purchase terms as a discount for early payment. The high variability of costs and availability has led to IO debt playing a smaller and smaller role in the farm operating debt market.

Farm borrowing from family and friends has also been declining. The size of the loans farmers need has grown as the sector has become more concentrated. Some farmers' loan requests now exceed the lending limits of local banks and are very likely much larger than individuals can provide. Increases in the size of loans not only exceed the ability of individuals to lend, but also increase the costs of record keeping and enforcement of payment. The integration of rural and national financial markets has also reduced individual lending by increasing the number of alternative investments available to the lender.

IO farm real estate debt has also shown a long-term decrease in importance. Since 1980, however, a moderate resurgence of seller mortgages has occurred as "creative financing" has been required to sell farmland in a period of historically high interest rates.

When, and if, interest rates decline the importance of seller financing of farm real estate may decline again. Farm real estate is a large portion of retiring farmers' assets. Thus the fortunes of retirees who finance the purchase of their land continue to be tied to risks of farming. Moreover, the continuing trend to fewer and larger farms suggests an increasing number of farmland transactions are occurring between unrelated individuals. Farmers are much more likely to finance purchases with their family members than with outsiders.

Over time, IO financing will likely play a smaller role in financing the farming sector. Debt will probably be provided through formal channels as the size of individual loans grows, competition among lenders increases and farming continues to become more of a business and less of a family life style.

Depository Institutions

Depository Institutions are primarily commercial banks. Other organizations, without bank charters, also accept deposits from individuals and businesses and either make loans or provide financing through purchasing bonds. Money market mutual funds are the largest and most widely recognized example of nonbank depository institutions. Brokerage houses do much the same thing, and, with the financial deregulation initiated in the early 1980s, many other large businesses are now offering financial services. In light of the revolution occurring in financial markets, and the potential role of nontraditional sources of debt to play a major role in agriculture in the future, discussions need to consider the lending activities of all depository institutions. Data sources and history, however, constrain descriptions of the roles of depository institutions in agricultural lending to the role of commercial banks. In the following discussion, therefore, the terms "banks" and "depository institutions" are used interchangeably.

Banks, particularly small rural banks, have found farmers to be good customers for the short-term loans needed to match the short maturities of bank deposits. Farm loans are usually invested in tangible assets that are identifiable and, under normal circumstances, grow in value. There are always risks in farm lending, however, even if customers are chosen carefully. Weather can destroy both crops and livestock without warning. Pest and diseases also pose problems although, in most cases, good management practices can minimize the likelihood of such problems having disastrous implications. Economic changes can also lead to significant loan problems when prices fall unexpectedly.

The timing of farm loan demand has caused problems for rural banks in the past. Seasonal and cyclical loan demands by all farmers or ranchers in an area increase at about the same time. Concurrently, farmers are also sending money out of the community to pay for inputs. In the past, when rural financial markets were isolated from other markets and available loan funds were limited by local deposits, the combination of high loan demand and lower levels of deposits occasionally produced shortages of loan funds.

The ability of small banks to service farm loan demand has also suffered from the growth in average farm size. Banking regulations limit the size of the loans that can be made to an individual borrower in order to limit the risk exposure of the bank in case of a default. As farm size has increased and the proportion of farm inputs purchased from nonfarm sources has grown, the size of some farm loan requests has exceeded the maximum size of loans available from small banks.

Several institutional arrangements have evolved to overcome periodic shortages of loan funds in rural markets and loan size problems. The existence of the Farm Credit System and particularly the Federal Intermediate Credit Banks can be traced, in part, to these imperfections in rural credit markets. Small rural banks have developed correspondent relationships with larger city banks. The Federal Reserve System set up seasonal borrowing privileges for its member banks that were subject to large swings in loan demand and deposit flows. Finally, groups of agricultural banks have joined together in different ways to directly tap national and even international money and capital markets.

Correspondent relationships are agreements between small rural banks and larger city banks that allow, among other things, the small banks to participate in loans with larger banks. Both banks are then considered to be making part of the loan and maximum loan size constraints are less burdensome. These relationships have been only partially successful in dealing with rural banks' problems, however. In normal times, larger banks have been able to provide funds to cover seasonal swings in rural loan demand. When money and credit growth has been constrained, the correspondent relationships have tended to break down. Larger banks service their own customers first, and when there is a general shortage of funds they reduce their participation in other banks' loans.

As part of the Federal Reserve System's responsibility to control growth in money and credit, banks are required to hold reserves against their deposits. Some banks in particular regions of the country have shown continuing periodic shortages of loan funds due to the cyclical nature of the industries they serve. Many of these areas service agriculture, but ski resorts and fishing villages face similar difficulties. To overcome these problems, the Federal Reserve System established the seasonal borrowing privilege for banking institutions that can document seasonal patterns of loan demand and deposit growth. Banks that are able to use the Federal Reserve System's Discount Window and satisfy requirements showing seasonal fluctuations are able to borrow reserves for the periods that they face shortages. The reserves allow banks to expand loans and increase deposits in their local area.

Finally, rural banks have banded together to sell farm loans to other organizations. By joining together, rural banks can put packages of farm loans together so that dollar values are large enough and risks are low enough to attract large institutional investors. In selling the loans, although continuing to service them, banks increase their loanable funds and decrease their risks of local production hazards. Some banks have created corporations that sell loans to the Federal Intermediate Credit Banks of the Farm Credit System and are generally known as Other Financial Institutions, (OFIs). Others have sold loans to large U.S. financial organizations that are big enough to raise funds on the national money markets. There is even one instance, where U.S. agricultural loans are sold to the Rabobank of the Netherlands (Barry). Thus, U.S. farmers and ranchers are now competing in international financial markets for loans.

Agricultural banks have, at times, had troubles satisfying their owners and their regulators as to the quality of their loan portfolios. About one-half of all farm loans are concentrated in about 4300 banks (Melichar 1984a). In these banks, farm loans account for more than 25 percent of the banks' total loans. The financial condition of these agricultural banks is strongly influenced by swings in the farm economy (Barry and Lee). Until 1984, however, agricultural banks have been conservatively managed and have had higher earnings and lower losses than other banks during economic downturns (Hughes). In 1984, the relative stability of agricultural and nonagricultural banks was reversed (Melichar, 1984b). There are concerns that regulators will not understand the cyclical nature of agriculture and the fragile markets for farm assets. If regulators force agricultural banks to foreclose on large numbers of farm loans in a short period of time the value of the assets serving as collateral for the loans can decline dramatically. Such declines could force more farm loans into problem categories and jeopardize the capital positions of the banks.

Concerns over the regulation of agricultural banks are compounded by the complexity of the regulatory superstructure of U.S. banking industry. Banks with a national charter are supervised by the Comptroller of the

Currency, part of the U.S. Treasury Department. Banks with state charters that have not joined the Federal Reserve System are subject to state regulations and supervision. State banks that are members of the Federal Reserve System and all bank holding companies are supervised by the Federal Reserve System. Moreover, any bank that carries federal deposit insurance is subject to the rules of, and examination by, the Federal Deposit Insurance Corporation. Given the number of regulators involved, strict adherence to rules and regulations is likely even at times when common sense might argue for a more flexible approach.

Banks generally lost market shares of farm debt to other farm lenders between 1973 and 1982. Part of the reason for this decline was banks' use of marginal pricing on loans and the short maturity of their sources of funds. Bank interest rates have, therefore, quickly responded to changes in market rates of interest. The Farm Credit System's use of average cost pricing of loans has generally put banks at a competitive disadvantage because interest rates trended upward over the 1970s. Other explanations of banks' loss of market share include their periodic shortages of loanable funds, their difficulties in meeting ever larger farm loan requests, growth in direct government lending to farmers and liberalization of regulations controlling Farm Credit System lending. Periodic shortages of funds and problems with being able to finance large loan requests have lead some farm borrowers to question the reliability of banks as a source of funds. Rapid increases in funds made available by government lenders at below market interest rates during the late 1970s and early 1980s decreased the market share of all private farm lenders. Finally, changes to Farm Credit System regulations have tended to reduce constraints on the System's growth and lead to a lower market share for all other lenders.

Life Insurance Companies

Life insurance companies have played a significant, but declining, role in farm real estate debt markets. In the mid-1950s loans from life insurance companies accounted for about 25 percent of all farm real estate debt. Since that time their market share has steadily declined reaching 11 percent at the end of 1983.

A natural match exists between the maturity of life insurance liabilities and long-term farm real estate loans as assets, similar to the match between the maturity of commercial bank deposits and farm operating loans. Thus, the failure of these companies to increase their farm loan portfolios as rapidly as farm debt has grown has been an interesting topic for debate. Insurance companies have shared some of the characteristics that are anticipated in a more integrated banking system. Once the commercial banking system becomes more integrated, rural banks will find they can invest their deposits in almost any sector of the economy, just as insurance companies have always been able to do. Thus, the reasons for the past decline in the importance of insurance company financing to the farm sector may provide some hints as to the future ability and desire of deregulated banks to compete for farm loans.

One of the principal reasons for the decline in life insurance companies' share of the farm real estate market is that the growth in their sources of funds has been slower than the growth in farm debt, particularly through the 1970s (Lins). During the 1970s incentives to save for the future, as indicated by real interest rates, were low. Thus, it was difficult to sell whole life insurance, which combined insurance against death with savings plans. Slow growth in life insurance premiums and fast growth in farm debt meant that an increasing fraction of the companies' investable funds were needed to keep their market share from declining. In fact, the opposite occurred.

Policy holder loans, usually carrying a fixed, contractually set interest rate, grew rapidly through the 1970s as market interest rates increased with growing inflation. These loans used a substantial portion of the companies' investable funds.

Net returns on farm real estate loans may have been lower than those available in other markets. Farm real estate is geographically disperse. Thus, the distribution system for farm real estate loans needs to be large, or significant travel expenses are incurred. The net returns on urban real estate loans or investments in secondary market mortgages may be greater while still matching maturities between assets and liabilities.

Finally, government involvement in farm lending may

have also led to interest rates on farm real estate loans that were lower than the returns available on loans to other sectors. If so, the future integration of the banking sector, which will open new investment alternatives for rural banks, may well lead to less bank financing for the sector.

Government Sources of Farm Debt

Three government organizations lend directly to farmers, the Farmers Home Administration (FmHA), the Commodity Credit Corporation (CCC), and the Small Business Administration (SBA). As of the end of 1984 FmHA was the largest direct government lender to farmers, holding 11 percent of total farm debt and the SBA was a distant third with less than one percent of total farm debt.

The organizations serve different purposes in the farm debt markets. Justifications of government interference in markets are normally based on market imperfections that exist without government involvement. Earlier in this century, the isolation of rural communities restricted the flow of financial resources to and from farming areas. Slow communications, an underdeveloped infrastructure and imperfect banking and commercial markets contributed to this isolation. Restrictions on the flows of financial resources were perceived as failures of financial markets, which provided justifications for direct government involvement in farm credit markets. For example, such arguments were used to help establish the Farm Credit System and later the Farmers Home Administration. The Commodity Credit Corporation was developed to help implement government commodity policies, and its lending to farmers is only a secondary issue to its purpose of supporting prices received by farmers. Thus, nonfinancial market failures provide the justification for the existence of the CCC, even though its programs provide a significant source of funding for farmers' inventories and have an impact on farm debt markets. The Small Business Administration exists, in part, to overcome the problems generated by the high risk of running a small business. In some cases, therefore, farms have qualified for SBA loans.

The Farmers Home Administration

FmHA is the primary rural credit agency of the Federal Government. It was created as part of the U.S. Department of Agriculture in 1946 to provide credit to rural residents unable to obtain credit from private lenders (Hughes, Gabriel, et. al.). FmHA took over and expanded the functions of the Farm Security Administration and the Rural Resettlement Administration. Since 1960, FmHA loans have grown by over 4000 percent. Most of the growth occurred in the 1970s when the scope of the organization's objectives experienced a major revision.

Loans are made in four basic programs: Farmer Programs, Rural Housing Programs, Community Programs and Industry Programs. Farmer Programs include: the Operating Loan Program, the Farm Ownership Program, the Soil and Water Program, the Emergency Disaster Loan Program and the Economic Emergency Loan Program. All programs are intended to serve only rural residents who are unable to obtain credit elsewhere. As such, FmHA was designed to be a "lender of last resort" to rural America, funding credit worthy projects that private lenders were either unable or unwilling to finance.

Originally, FmHA operating loans and farm ownership loans were developed to help beginning farmers who did not qualify for private financing because of inadequate collateral or due to a lack of experience with large amounts of debt. The Emergency Disaster Program provided funds for farmers who had sustained a significant loss due to natural disasters beyond their control. Starting in farming and the setbacks from natural disasters were considered to be temporary conditions and farmers were expected to "graduate" from these programs to private credit within a specified number of years. Soil and water loans facilitated long term improvements to the land which were in best interest of society. Such projects may never have generated sufficient returns to private landowners to be undertaken with market interest rates.

In 1978, a new farm program was introduced entitled the Economic Emergency Loan Program. This new program authorized FmHA to make loans to farmers when credit was scarce due to monetary policy, low product prices or high production costs. While on the surface this new program was similar to existing programs, its implementation brought a whole new role to FmHA. Instead of overcoming temporary market failures by lending to farmers who would eventually graduate to private credit, this new program put the FmHA in the position of stopping sector level adjustments to changes in the economy. If, for example, a new technology changed the supply of food so prices fell, all farmers who were unable to obtain credit from private lenders would qualify for FmHA financing. If the shift to lower prices was permanent, those farmers might never be expected to graduate. As the implications of this program became evident, the program's funding has been reduced and few new loans have been made since 1982.

Overall, the Farmers Home Administration serves many valuable functions: overcoming the shortcomings of the financial markets, dealing with the risks of beginning farmers, encouraging soil and water conservation projects, and to some extent offsetting the impacts of natural disasters. Much of the criticism of FmHA in recent years has centered around the Economic Emergency Loan Program. This program, which had a weak economic justification, has reduced the ability of the sector to adjust gradually to changing economic conditions and may well have exacerbated current farm credit problems.

The Commodity Credit Corporation

The Commodity Credit Corporation originated in 1933 as part of a set of farm programs designed to support and stabilize farm commodity prices (Hottel). The CCC provides nonrecourse loans to eligible farmers and farmer cooperatives to help finance orderly marketing and support the prices of targeted crops. A nonrecourse loan is one that can be repaid either by paying back principal and interest or forfeiting collateral. Since crops are the collateral for these loans, the amount loaned serves as a minimum farmers will have to accept for their production. If market prices are too low, the farmer forfeits the crop to the CCC and keeps the money from the loan. If prices are sufficiently high, they sell their crops and repay the loan in a normal fashion.

Two types of programs to support farm prices and income are currently in use, the regular nonrecourse loan program and the farmer owned reserve program. For most major crops, the program provides loans for up to nine months to help farmers distribute their marketings through the year. The farmer owned reserve program can be entered at the end of the normal loan period. When entering the reserve program the farmer agrees to keep the crop in storage for three years, or until the price of the crop rises above a predetermined release price.

The CCC also makes loans to finance farm storage facilities. These loans provide for the construction of structures that give farmers additional flexibility in marketing their production. During the large surpluses of the early 1980s these loans grew rapidly as commercial storage facilities filled and farmers needed to build their own storage to protect their crops from deteriorating.

The amount of CCC debt outstanding and its importance in farm lending are, therefore, determined by several factors. Included are the current prices for targeted crops, the amount the CCC will loan on each unit of production (the loan rate), interest rates from other lenders, and the fraction of production eligible to be placed under loan, which is itself determined by what portion of farmers participate in the other conditions of farm commodity programs such as acreage set asides. At the end of 1981, CCC held eight percent of farm nonreal estate debt, in 1982, 14 percent, and in 1983, ten percent. The growth during 1982 was due to large increases in commodity inventories while the decline in 1983 was due to the distribution of CCC assets as part of the Payment in Kind Program.

The Small Business Administration

The Small Business Administration was created in 1953 as an independent government agency designed to provide credit to small businesses unable to obtain credit from the private sector. Because FmHA serves a similar function and has more expertise in dealing with farm loans, no funds were lent to farmers until 1976 when the SBA was forced to consider farmers' requests by a Congressional mandate. Farm loans from the SBA peaked at less than $2 billion in 1978. Congress then ruled that farmers must first apply to the Farmers Home Administration before applying to the SBA. Since that time few, if any, new farm loans have been made by the SBA.

Leasing As an Alternative to Debt

Two types of assets are generally leased in farming, land and machinery. There are, of course, exceptions to this including the leasing of everything from storage sheds to pregnant cows. Land is by far the most valuable asset on farm balance sheets, being valued at $764.5 billion at the end of 1984. Machinery and equipment is the second largest category at $111 billion. The following discussion, therefore, concentrates on land and machinery leasing.

Lease financing provides certain benefits over debt financing even though its impact on a business's financial condition may be similar (Adair, et. al.). Lease financing can, in some cases, be accomplished with no equivalent to a downpayment. This might be the only way a business can achieve a 100 percent leverage position for an asset. Leases may have longer maturities than loans available for financing the same asset. Tax advantages in leasing may occur if the lessor has a higher tax bracket and can use the depreciation and investment tax credit to more advantage than the lessee. Finally, leases may reduce the lessees' risk exposure for assets with high risks of obsolescence.

Land Leases

There are two standard types of farmland leases. Cash leases require the lessee, or tenant, to pay a fixed dollar amount per year for use of the land. Share leases, both crop and livestock, require the tenant to pay the landlord with a prespecified fraction of the output produced on the land.

The principal difference between the two types of leases is in which party bears the risk of production. In cash leases, the rent paid is a fixed business expense to tenants and any fluctuations in yields or prices directly affect their net income. The lessor is totally insulated from the risks of farming, unless the tenant goes broke and cannot pay the rent.

In a share lease, the owner usually shares in the operating expenses, as well as providing the land. The tenant and landlord then split the resulting output, be it crops or livestock, at the end of the growing cycle. Risks of changes in both the quantity and value of production are shared by both parties.

The effects of land leases on sector financial statements are controversial. The Balance Sheet of the Farming Sector includes all land in farms regardless of ownership. It also includes all debts that are collateralized by farm real estate. Other categories of assets and liabilities, however, reflect only those owned and owed by farm operators and their families. Confusing the issue even further, particularly when financial ratio analyses are done, such as calculating rates of return, is the fact that rents to landlords are subtracted from net farm income. Thus, the net returns to farm operators are usually divided by the value of assets of both operators and landlords. This understates rates of return in the sector and may be used to justify excessive government involvement.

On an individual farm basis, the confusion of accounting information can be even greater. Farm accounting standards are not, as yet, generally practiced and leases are in many cases not reflected in the balance sheet. The payments contracted in a lease, particularly a long-term lease, present claims on future returns similar to those of long-term debt. The sale of an asset with a concurrent long-term lease-back can, therefore, be used to make cosmetic improvements to farm businesses' financial statements.

Leasing Machinery and Equipment

There are also two principal types of leases used in leasing machinery and equipment: operating leases and financial leases. An operating lease for a piece of equipment involves the use of the equipment for a short period of time. This type of lease is usually used to cover emergency situations when an important piece of equipment breaks down during planting or harvesting.

In terms of dollar volume, financial leases are probably more important. Financial leases are defined to be "a noncancellable contractual commitment for the lessee to make a series of payments to the lessor for the use of an asset, although the lessor retains title to the asset" (Adair, et.al.). Survey results showed that financial leases to farmers in 1980 totaled about $0.6 billion. Survey respondents had seen their leases grow by 140% since 1979 and expected another 234% increase by

1985.

Growth between 1980 and 1983 may have exceeded expectations due to the 1981 Economic Recovery Tax Act. This law reduced constraints on the definition of a lease for tax purposes. As a result, financial leases in the economy grew dramatically to transfer tax deductible depreciation to businesses with the highest tax brackets. In 1982, however, the Tax Equity and Fiscal Responsibility Act reversed the changes made in 1981, and by the end of 1983 all leases were required to pass tests similar to those that existed prior to 1981 (Durst, et. al).

Unfortunately, farm sector accounts contain even less information about leases of machinery and equipment than they do about land leases. Because there is no mention of these leases, assets and liabilities are understated, income is overstated and cash flow information is distorted. Reporting these leases in individual firm accounting statements suffers from the same problems as do land leases.

REFERENCES

Adair, Ann Laing, John B. Penson, Jr., and Marvin Duncan, "Monitoring Lease-Financing in Agriculture," Economic Review, Federal Reserve Bank of Kansas City, June, (1981) pp: 16-27.

Barry, Peter J., "Innovative Funding of Farm Loans by Commercial Banks: The MASI Experience," Illinois Agricultural Economics Staff Paper, No. 83-E267, August, 1983.

Barry, Peter J. and Warren F. Lee, "Financial Stress in Agriculture: Implications for Agricultural Lenders," American Journal of Agricultural Economics, Volume 65, December, (1983) pp: 945-952.

Durst, Ron, Wendy Rome and James Hrubovcak, "The Economic Recovery Tax Act of 1981: Provisions of Significance to Agriculture," USDA, ERS, NED, Staff Report, AGES 810908, September 1981.

Hottel, Bruce, "The Commodity Credit Corporation and Agricultural Lending," Agricultural Finance Review, USDA, ESS, Volume 41, July, (1981) pp: 73-82.

Hughes, Dean W., Stephen C. Gabriel, Ronald L. Meekhof, Michael S. Boehlje and George R. Amols, "Financing the Farm Sector in the 1980's: Aggregate Needs and the Roles of Public and Private Institutions," USDA, ERS, NED, Staff Report No. AGES 820128, February, 1982.

Hughes, Dean W., "Financial Condition of Agricultural Lenders in a Time of Farm Distress," Economic Review, Federal Reserve Bank of Kansas City, July-August, (1983) pp: 13-31.

Lins, David A., "Life Insurance Company Lending to Agriculture,: Agricultural Finance Review, USDA, ESS, Volume 41, July, (1981) pp: 41-49.

Melichar, Emanuel, "A Financial Perspective on Agriculture," Federal Reserve Bulletin, Board of Governors of the Federal Reserve System, January, 1984a.

Melichar, Emanuel, "An Overview of Agricultural Banking Experience," Proceedings Financial Stress in Agriculture Workshop, Kansas City Missouri, October 22, 1984b.

U.S. Department of Agriculture, Economics Indicators of the Farm Sector, Insome and Balance Sheet Statistics, Economics Research Service, various years.

5

Projecting Financial Conditions in the Farm Sector

Dean W. Hughes

The purposes of this chapter are twofold. One is to describe the tools (models) available to farm policy-makers and industry decision-makers to project future financial conditions in the farm sector. The other is to link previous chapters dealing with historic problems to chapters describing future issues in farm finance by providing an example set of projections. Analysis of probable future conditions is important to policy-makers in designing policies that do not simply react to past problems, but are intended to properly interact with the sector in coming years. Industry decision-makers are perhaps even more interested in the future, since the profitability and even the economic survival of their organizations depend on making appropriate decisions on investments that last for many years.

It is difficult to define what constitutes a financial model. Almost all estimates of the future of agriculture have a financial component and, with few exceptions, have financial implications. For example, knowledge of recent levels of farm income is generally required to identify likely future production and marketing decisions. Information on asset quantities and prices is also important. Moreover, the results of almost every prediction of what agriculture will be, or will do, in the future have an influence on the sector's income statement and balance sheet. It is, therefore, difficult to make any sharp distinctions between financial and nonfinancial projections. In the broadest sense, all statements about the future of agriculture are projections of future financial conditions. However, only models that include information about at least sector income will be considered as agricultural financial projection models in the rest of this chapter.

A distinct difference exists between the words projection and forecast. Projections provide answers to "what if" questions. Forecasts are statements about "what will be". Economic analyses are constrained to deal with projections since forecasts require information beyond the realm of economic study. Political, technological, weather and sociological factors probably play as important a role as economics in determining what will actually happen in the world. Thus, economists are limited to describing only the economic ramifications of future events. The normal tendency of those not familiar with these limitations is to interpret projections as forecasts. This has been detrimental to the profession, and at times even generated questions about the justification of economic studies.

Projections are conditioned by the assumptions made about external forces as well as the economic relationships included in the analysis. It is, therefore, important that the users of economic projections understand these assumptions. Also important is that economists understand the limitations of their work and clarify these limitations in the communication of their results.

Types of Projections

Methods used in making projections are usually divided into two categories, subjective and objective. Subjective estimates are the most common and are made by almost all participants in the farm sector. Anyone involved in the sector has opinions about the future of agriculture. Perhaps they cannot, or will not, put numbers to their estimates, but they are usually willing to say things are going to get better or worse. When pressed, they normally agree that better means higher incomes and rising asset values, while worse implies reduced incomes and declining asset values. Objective estimates of future conditions are normally associated with mathematical models of parts of the economy which generate specific numbers on prices, production, incomes, and asset values. In many cases, these models are solved using computers and results are presented in tabular or graphical forms.

Subjective and objective estimates of future conditions are far from distinct when examined closely. Subjective estimates are usually the distillation of many inputs: the estimator's current economic condition, conversations with colleagues, presentation's by "experts", and the reading of newspapers, trade journals, and other reports. Objective estimates form the basis of many of the experts' presentations and are often reported widely in the newspapers, trade journals, and reports. Thus "objective" estimates form at least part of the basis for "subjective" estimates. On the other hand, there are no truly "objective" estimates of future conditions. Economic models are simplications of reality. Depending on how comprehensive they are, such models, therefore, require subjective estimates for at least some of the following data: future policy decisions, international economic conditions, and domestic nonagricultural conditions. In models of small scope, for example a model of the poultry subsector, estimates of feed prices might be required which clearly incorporate subjective evaluations of future conditions in other parts of agriculture. Thus, subjective impressions are incorporated in objective projections and the quality of the projection is only as good as the knowledge of the person using the model.

The advantages of objective models are threefold. Objective models, with their explicit connection of one result to another, produce projections that are generally more consistent than subjective projections. Many of the models used to project financial conditions in the farm sector have literally hundreds of equations that connect such things as income to investment, and investment to the level of production. Subjective projections can disregard some of these linkages and be internally inconsistent. And, inconsistent projections can be biased by emotions, reducing the need to face the possibility of difficult choices. Secondly, objective models allow counterfactual analyses to identify the impacts of different decisions. For example, models can be run with different levels of commodity loan rates to see how much they effect farm incomes. Finally, objective models usually provide a starting point for new subjective projections. Debates over new farm policies normally generate many objective projections of the possible impacts of proposed changes. While few people can, or should, accept these projections as exactly correct, most accept them as a place to begin discussions in the policy-making process.

Objective Models

The development of models capable of projecting sector financial conditions has been tied to advances in computer technology. Prior to computers, identifying the reaction of one economic variable to a change in another was a time consuming process. And, the solving many equations at the same time to project future outcomes often exceeded the time needed for actual changes to occur in the economy.

Once introduced, however, models of the farm sector have gone through three generations (Hughes and Penson 1981). First generation models treated the farm sector as a separate entity. In those models, projections were based on subjective evaluations of the future course of domestic economic conditions, such as, inflation, interest rates, and total incomes which significantly affect conditions in agriculture. The second generation of sector models continued to treat agriculture as separate from the rest of the economy, but tied sector projections to the results of running objective models of the whole economy. Implicit in second generation models, however, was the assumption that the future of agriculture has no impact on the rest of the economy. To overcome the problems generated by this assumption, a third generation of models is currently being developed that solve for general economic and agricultural outcomes at the same time. In these models, conditions in the nonagricultural and agricultural sectors interact, and results are therefore consistent between the two.

All three generations of models remain in use. The USDA, for example, has a policy model, called FAPSIM, that is a first generation model and is used extensively. While it requires prespecifications of general economic conditions, its farm commodity detail is so extensive that is is capable of answering some questions other generations of models are not, as yet, capable of addressing. Second generation models are provided by such well known names as Chase Econometrics and Wharton Econometric Forecasting Associates. Large, respected general economic models are used to generate results that form the assumptions of the future economic environment for agricultural sector models. Third generation models are generally located in universities. One such model is currently being used at Texas A&M and Texas Tech Universities, and one is being developed at the University of California at Berkeley.

Of the third generation models, only one, COMGEM at Texas A&M and Texas Tech, was developed specifically to project financial conditions for the sector. In addition to incorporating agriculture as part of the economy, the model keeps track of farmers' incomes, assets, debts, and cash flows. This information is linked within the model to projections of farmers' production and marketing decisions, as well as consumers' food purchases and other parts of the economy as a whole. The model is used by the U.S. Department of Agriculture in some of its policy analyses and forecasting activities.

Projections of Future Financial Conditions in the Farm Sector

The following projections were made using COMGEM version 80.2 in the fall of 1984 (Penson, et. al.; and Hughes and Penson 1985). At that time the size of the financial problems for farmers in the 1980s was becoming generally recognized and preparations for the 1985 Farm Bill were underway. Their original intent was to highlight the importance of macroeconomic policies in determining farmers' financial situation and provide policy-makers with a framework for their discussions of the benefits of debt relief legislation.

Three alternatives are presented below to reflect projections of possible courses of monetary and fiscal policies. The reader of this book will, of course, have the benefit of having lived through at least some of the periods projected. With the benefit of hindsight, the inaccuracies of the projections will be self-evident. Yet, the purpose of presenting this information is not to perfectly predict the future, but to provide the reader with background information on many of the issues arising in agricultural finance and tax policies and why they are important.

Three projections are presented: S1, S2, and S3. S1 assumes continued high government deficits through 1990 and a renewed fight against inflation using a restrictive monetary policy. Inflation in the mid-1980s was still between 4 and 5 percent, levels almost as high as those that generated price and wage controls in the 1970s. S2 also assumes high deficits, but a more relaxed monetary policy. Past history would suggest continued cycles of low growth and low inflation followed by rapid growth and higher inflation. Moreover, slow money growth in the face of large deficits might be interrupted by concerns over a collapse of international financial markets. S3 assumes decreases in the deficit and a monetary policy between that used in S1 and S2. As such, S3 represents an optimistic view of the future.

No attempt is made to outguess farm policy-makers in any of the scenarios. Farm policies are the same across scenarios and reflect 1984 levels of support without the delayed impact of the 1983 Payment In Kind program. No surprises in either energy prices or weather are assumed in any of the projections. All results include as much of the information on farm households as is normally available from the USDA.

S1 - High Deficits and Slow Growth in Money

In this scenario, the real government deficit is held at its 1984 level through 1990. Monetary policy is constructed to eliminate inflation by 1987, and then, contain changes in the overall price level to almost zero. It is, of course, highly unlikely that such a course could actually be achieved. Eliminating inflation would probably introduce more volatility in the economy than shown in the model results. However, the implications for the farm sector of another period of disinflation in the face of continued large deficits are clearly indicated in the model results.

Stopping inflation in the face of continued fiscal stimulus is projected to lead to another period of slow growth and higher real interest rates for the whole economy. The solid lines in Figures 1, 2, and 3 show the percent change in real GNP, the percent change in the GNP deflator, and the real prime interest rates for this scenario. Growth in real GNP, which averages 2.3 percent from 1985 through 1990, is substantially below its long-run potential growth rate of about three percent. Inflation is smoothly controlled by 1987 and after that the GNP deflator does not change by more than 0.2 percent annually. The real prime interest rate continues to grow

FIGURE 5.1: PERCENT CHANGE IN REAL GNP

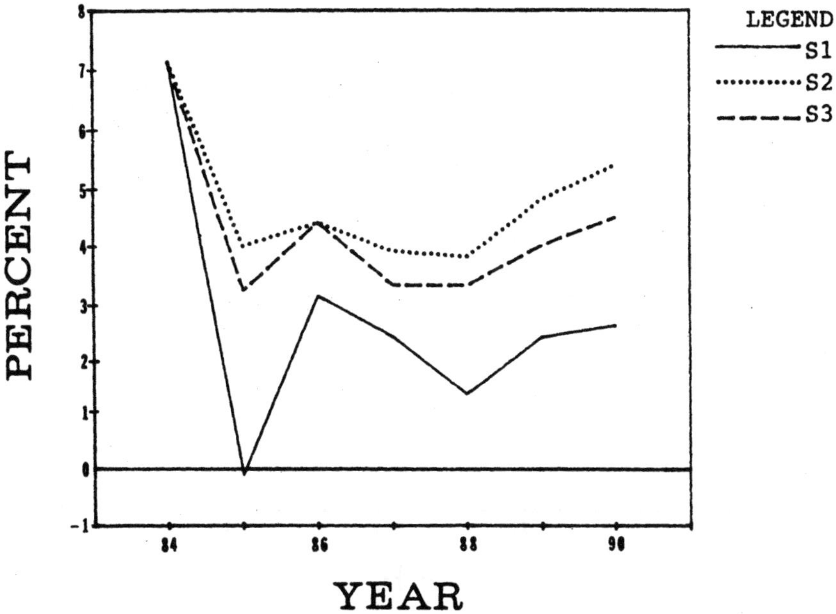

FIGURE 5.2: PERCENT CHANGE IN GNP DEFLATOR

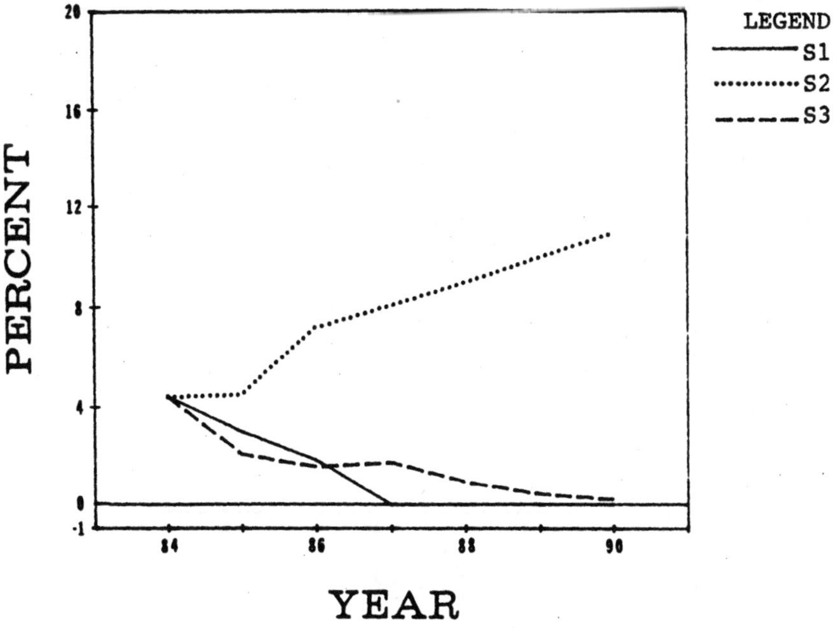

FIGURE 5.3: REAL PRIME INTEREST RATE

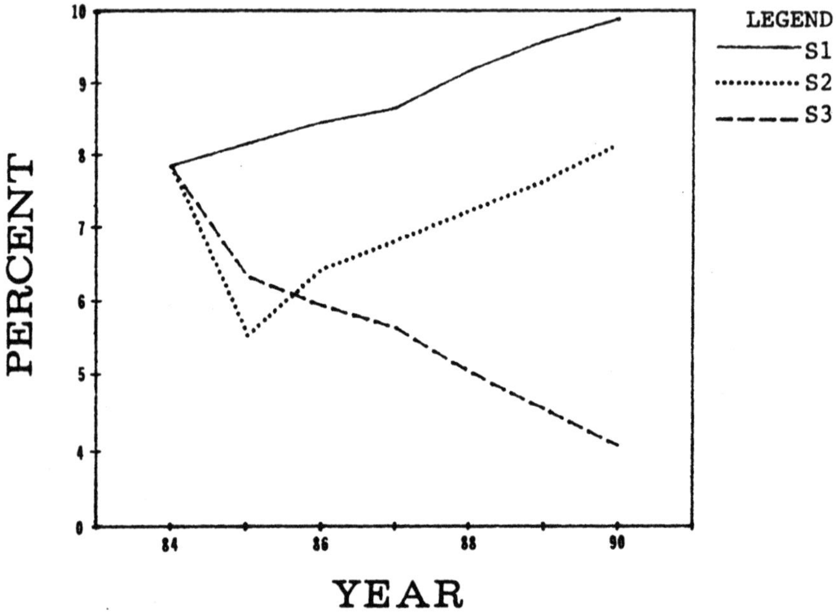

FIGURE 5.4: REAL NET FARM INCOME

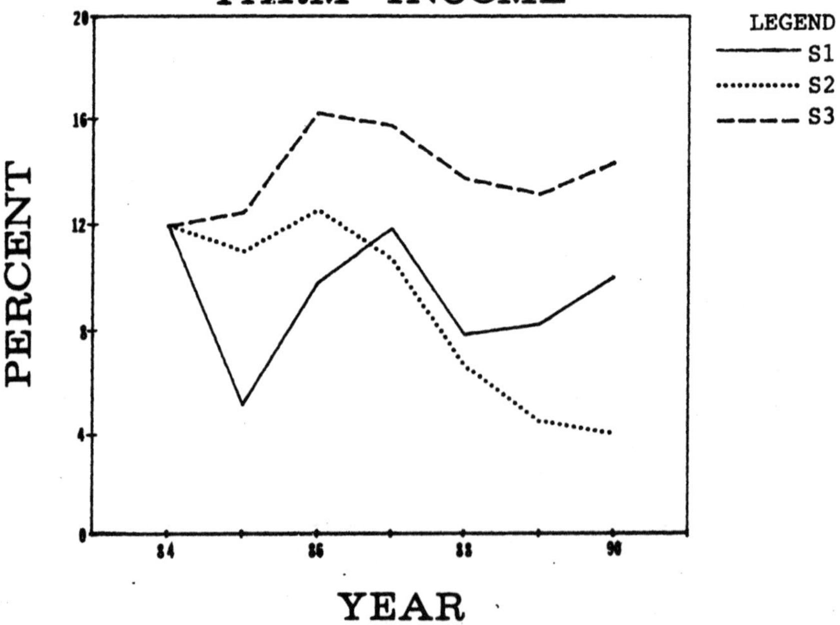

FIGURE 5.5: REAL FARM ASSET VALUES

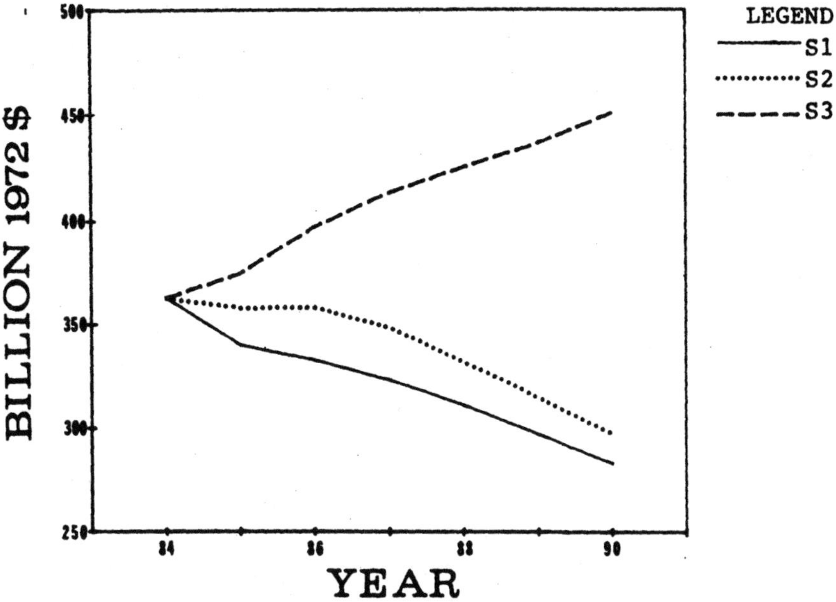

FIGURE 5.6: REAL TOTAL FARM DEBT

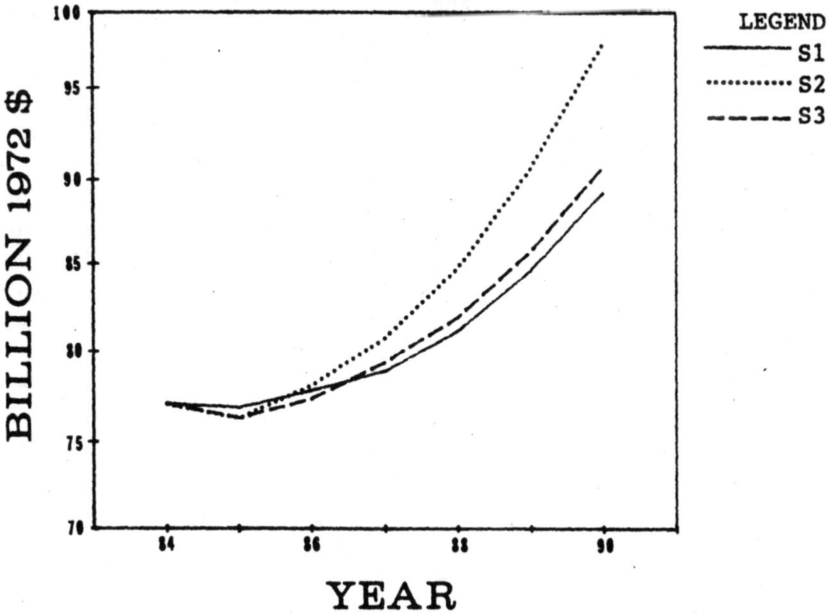

throughout the period, reflecting both the continuing struggle between fiscal and monetary policy and bank adjustments to an environment with higher cost funds.

The implications for the farm sector of continued slow growth in incomes and higher interest rates are obvious: low incomes, declining asset values, and restricted use of debt financing. Figures 4, 5, and 6 graphically illustrate these points with the solid lines again representing results from the first scenario. In 1967 dollars, net farm income fluctuates between $5 and $12 billion with no apparent trend. Its average for the decade 1981-90 would then be $9 billion compared to $15 billion for 1971-80.

Farm asset values decline substantially in this scenario. By 1990, they might be 82 percent of their 1983 value in nominal terms and 75 percent in constant dollar terms. All of the decline is in physical assets with the principal loss coming from farm real estate values. Financial assets grow, reflecting the incentives for participants in the sector to divert their available funds into nonfarm assets.

Debt is projected to grow, but at a much slower rate than in the 1970s. The average growth in nominal farm debt for this scenario during the 1980s would be 4 percent versus 13 percent in the 1970s. Almost all of the growth is in real estate debt. While real estate is losing value in the projections, it still provides the safest collateral for farm loans. Moreover, low net incomes and likely repayment difficulties would probably lead to refinancing short-term debt over longer repayment periods. Such financing would almost certainly shift debt to the real estate category.

The net results of declining assets and increasing debt are a substantial decline in sector wealth and an increase in leverage. Proprietors' equities decline 29 percent in nominal terms and 35 percent in constant dollars. The debt-to-asset ratio for the sector grows from 21 percent in 1983 to 32 percent in 1990. Certainly, fewer farms would be debt-free if these projections are realized. In addition, farmers with low debt-to-asset ratios in 1983 would be highly leveraged by 1990. And, those farmers who currently have moderate to high leverage ratios will likely be out of business.

The sector's cash sources and uses of funds statement would also reflect the effect of S1's macroeconomic policies. Capital purchases by sector participants to maintain equipment and structures and buy the equity of discontinuing proprietors averages 79 percent of net cash income between 1985 and 1990 versus 57 percent in 1971 through 1979.

The principal reason for this shift is the decline in the net inflow of debt funds. The ratio of the net flow of loans to total capital flow declines from an average of 48 percent in the 1970s to a projected average of 16 percent from 1985 through 1990.

Prospects for the sector are dim if the deficit remains high for the next 6 years, and further reductions in inflation are achieved exclusively through monetary policy. Farm incomes are likely to remain low, asset values will probably continue to decline, and debt financing might be constrained. Under these conditions, the sector would probably end the decade in very weak financial health. It would be unable to show much resilience to change and perhaps be less productive.

S2 - High Deficits and Faster Growth in Money

In this scenario, the real government deficit is again held at its 1984 level through 1990. Monetary policy is assumed to allow the inflation rate to grow at about one percentage point per year. The impact of these policies is principally shown in real interest rates that first decline in 1985 and then return to 1984 levels by 1990 (dashed lines in Figure 3). The surprise of having monetary authorities reverse their policies is projected to allow both nominal and real rates to decline early in the projection period. As the new policy becomes generally accepted and built into expectations, however, real rates grow to clear financial markets still servicing huge government borrowing demands.

Growth in real GNP (dashed lines in Figure 1) is higher than rates of growth in S1 for all years. Since the growth is higher than the sustainable three percent increase in potential GNP, such growth is probably being borrowed from future periods. Sometime after 1990, a renewed fight against inflation will likely be needed accompanied by a recession or recessions that will reduce GNP below its long-run growth path. Inflation (dashed lines in Figure 2) is held low through 1985 by policies already implemented during 1984. It is projected to then jump about 1.5 percentage points in 1986 and grow at a rate of about one percentage point per year through 1990.

Projections of the impacts of this set of macroeconomic policies on the farm sector are less obvious, but perhaps more interesting than those for the first scenario. Basically, the projections suggest that faster growth in money helps the farm sector in the short-run (about 3 years), but harms the sector over the long-run. Real net farm income (dashed lines in Figure 4) is projected to be higher than in the first scenario through 1986 due principally to higher domestic and foreign demands for food. The recession projected for the domestic economy in 1985 under S1 is not expected with an easier monetary policy. The early reduction in U.S. real interest rates in this scenario decreases the value of the dollar relative to other currencies and stimulates demand for agricultural exports. By 1990, however, the growth in real interest rates increases the value of the dollar, decreases export demand, and adds to farmers' interest expenses. Moreover, the high rate of inflation in this scenario generates rapid increases in farm production expenses. Long-run constraints on growth in demand and rapid increases in production expenses generate projections of real net farm income in the late 1980s that are substantially below those of alternative scenarios.

The decline in real farm asset values (dashed lines Figure 5) with faster growth in money is slower in the first few years of the projections, but faster in later years because farm profitability is lower. One point to note is that due to rising inflation, projections of nominal farm asset values increase throughout the projection horizon, although at much slower rates than were realized in the 1970s.

Growth in real farm debt is highest for this scenario, although still slower than the rates in the 1970s (dashed lines in Figure 6). The average annual compound growth rate in real debt from 1984 through 1990 in S2 is four percent compared to six percent in the 1970s. The reasons for the faster growth in debt are related to the farm debt repayment problems caused by the declining incomes after 1986, lower real interest rates, and slower declines in asset values when compared to the first scenario. These generate more demand for debt financing and provide somewhat more collateral to allow the demand to be realized.

The sector cash flow statement and financial ratios for S2 confirm the stories told by the income statement and the balance sheet. After 2 or 3 years of improved financial conditions, the impacts of more rapid inflation on the sector are evident. Capital purchases start to exceed net cash income from farming in 1989. More loan funds are used in the late 1980s than in the first scenario. While loan funds are projected to account for an average of only 27 percent of purchased capital in 1984 through 1986, the average increases to 77 percent in 1987 through 1990. The ratio of debt outstanding to total cash income grows from 3.8 in 1983 to 6.6 in 1990. Finally, the debt-to-asset ratio for the sector increases from 21 percent in 1983 to 33 percent in 1990. Such an increase indicates that many of the same adjustments expected in the first scenario would be required, even with a less restrictive monetary policy. The principal difference between S1 and S2 is that, when problems are postponed by inflating the economy, adjustments are made in a shorter period of time.

Monetizing the deficit is, therefore, probably not the solution to farm sector problems. It leaves the farm sector at the end of the decade with lower incomes, slightly higher real estate values, and more debt than if monetary policy is restrictive.

S3 - Lower Deficits and Moderate Growth in Money

In this scenario, the real government deficit is assumed to decline by about 15 percent per year through 1990 due to an imposed slowing of the growth in government expenditures. Monetary policy is one of moderate growth in money such that inflation is less than one percent by the end of the decade. Again, it is hard to imagine economic policy-makers being able to generate as smooth a transition to a noninflationary economy as is shown by the model. The implications for the farm sector of such a transition are, however, enlightening.

The projection of growth in real GNP lies between the projections of the other scenarios. It is higher than the first scenario because monetary policy is less restrictive. It is lower than that of the second scenario, because monetary policy is more restrictive and fiscal policy is less stimulative.

Inflation (dotted lines in Figure 2) is the driving force in developing monetary policy for this scenario and, therefore, rather smoothly declines to almost zero in 1990. It is able to drop more quickly in 1985 and 1986 than inflation in the first scenario due to the reduction in fiscal stimulus to the economy. Its rate of descent is less after 1986, reflecting a less severely restrictive monetary policy.

Real interest rates (dotted line in Figure 3) show the impressive reductions that can be gained when monetary and fiscal policies cooperate in reaching a goal. Continued reductions in government borrowing reduce demand for loanable funds. On the other side of the market, more moderate restrictions on the growth in money and credit add to the supply of loanable funds. The net result is declining real interest rates throughout the period. Moreover, declines in inflation help nominal interest rates to fall precipitously.

Here again, the implications for the farm sector are clear: higher incomes, rising asset values, and moderate growth in the use of debt financing. Real net farm incomes (dotted line in Figure 4) rise rapidly and then fluctuate at higher levels. While projections of real incomes do not reach the level of 1973, which was $26 billion, they substantially exceed those of the early 1980s. The projected average real net farm income for 1985 through 1990 is $14 billion compared to $9 billion for 1980 through 1983. Reasons for this increase lie in the change in interest and inflation rates. Lower interest rates decrease the value of the dollar and expand export demands. They also help increase domestic demand by stimulating economic growth and reducing inventory carrying costs. At the same time, lower rates of inflation keep production expenses from growing rapidly. The net result is a higher level of profitability for the sector.

When higher profitability is combined with lower interest rates, farm asset values start to increase again (dotted line in Figure 5). The average annual compound growth rate for real farm asset values from 1985 through 1990 for this scenario is four percent. While this is less than the six percent real growth in the 1970s, real interest rates in the scenario continue higher than those in the last decade. Most of the gain in the value of assets is in farm real estate, although machinery and equipment also show steady gains.

Almost surprisingly, real farm debt (dotted line in Figure 6) is projected to grow moderately in this scenario, a 3.5 percent average annual compound growth rate from 1985 through 1990. The reasons for this slow growth are that more internal funds are available due to higher levels of profitability, and real interest rates, while declining, continue to be high by historic standards.

The cash sources and uses of funds statements and financial ratios show the farm sector in much better financial condition by the end of the decade. Purchases of capital make up only about two-thirds of net cash income. And, the ratio of debt outstanding to total cash income declines from 3.8 in 1983 to 3.0 in 1990. The sector's debt-to-asset ratio also declines from 21 percent in 1983 to 20 percent in 1990. Under the macroeconomic policies assumed in this scenario, adjustments currently underway in the sector will likely be stopped or reversed. Instead of going out of business, highly leveraged farmers and ranchers would likely do very well. Farmers who prefer to be without debt could probably avoid borrowing. And, farmers with moderate debt-to-asset ratios would not be forced into more highly leveraged situations.

Contrary to the results of other scenarios, reductions in government deficits and moderate restraint in monetary policy could lead the farm sector to an improved financial condition by the end of the decade. Steady economic growth, lower inflation and lower interest rates would probably result in higher farm incomes and asset values, and moderate growth in farm debt. This would, in turn, mean the sector could be more resilient to weather and other shocks and be more productive than it is today.

General Implications for Farm Credit Policies

Unless there are substantial changes in fiscal policy, farm policy or weather conditions, the prospects for the farm sector are dim, at least through the end of the 1980s. Incomes, asset values, and equities may well decline through 1990. Growth in debt is likely to come from an inability to repay old debt and insufficient internal funds to meet desired cash uses.

Such a situation would have definite implications regarding what organizations will finance the sector.

Banks and insurance companies, which have alternative investment possibilities, would likely turn away from the sector. The Farm Credit System would be faced with higher interest rates on its bonds to cover the risk premiums attached to investors' concerns over conditions in agriculture. With the resulting higher interest rates on its loans, fewer farmers would be creditworthy. The government's role in farm credit markets would likely expand. Pressures would build for additional funding for the Farmers Home Administration as more and more borrowers become unable to obtain credit from private lenders. Increasing numbers of farmers would probably participate in commodity programs and be eligible for Commodity Credit Corporation loans.

Changes in the sources of farm equity are also likely to result. Continued problems in agriculture without problems in the general economy may lead to significant increases in land rentals from nonoperator landlords. Large farms may well develop new ways of tapping equity funds, perhaps by growing large enough to sell stock. Leasing would also become more popular if the sector continues in recession, because tax rates would be higher at nonfarm firms and large downpayments for farm machinery and equipment would be unavailable.

Reductions in government deficits coupled with moderate restraint in the growth of money and credit would likely produce different policy implications. A profitable sector would bring about increased competition in the farm credit markets. Banks would probably aggressively pursue farm loans and small rural banks would suddenly find themselves as takeover targets by larger banks. Savings banks might start to compete for farm real estate loans. Insurance companies would likely expand their farm loan portfolios. The Farm Credit System could find itself at a competitive disadvantage if it continues to price its loans at its average cost of funds and interest rates generally decline over the period. Direct government lending might decline and perhaps return to a goal of simply trying to overcome market imperfections in dealing with risk. Finally, leasing would likely show less growth if the benefits of ownership could be captured directly by farmers and cash flow is less of a constraint.

REFERENCES

Hughes, Dean W. and John B. Penson, Jr., "The Value of Endogenizing Agriculture in a MultiSector Macroeconomic Model," RWP 81-07, Federal Reserve Bank of Kansas City, August, 1981.

Hughes, Dean W. and John B. Penson, Jr., "Effects of Selected Macroecnomics Policies on Agriculture: 1984-1990," Agricultural Finance Review, Volume 45, 1985.

Penson, John B. Jr., Dean W. Hughes and Robert F. Romain, "An Overview of COMGEM: A Macroeconomic Model Emphasizing Agriculture," Departmental Information Report, DIR 84-1, SP-12, Texas Agricultural Experiment Station, Texas A&M University, December, 1984.

6

Current and Future Issues in Agricultural Capital Markets

Peter J. Barry

This chapter considers the emerging issues in agricultural capital markets that will shape the financing of U.S. agriculture in the later 1980s and beyond. First, a brief review of the relevant background of these financial markets and the evolution of credit conditions in agriculture up to the mid-1980s is presented. Then primary focus is given to the regulatory and policy issues affecting the major commercial farm lenders: the commercial banking industry and the Cooperative Farm Credit System (FCS). The prospects for continued development of leasing and equity financing are also addressed.

These issues are timely because the beginning of the 1980s has witnessed a number of landmark events in agricultural finance that will have lasting consequences. In the financial markets, the wide swings in interest rates and significant deregulation of depository institutions more closely tied farm credit to national and international markets, and made the farm sector more susceptible to changes in world economic conditions. Other financial market issues specific to agriculture involve concerns about the scope and funding of the FCS, the proper role of public credit programs, and the status of outside equity capital as a source of financing for farm businesses. In agriculture, farm income swung widely during the 1970s and 1980s creating one of the deepest and most widespread recessions in farming since the 1930s. Many farmers who used substantial debt capital experienced severe financial stress. Increasing pressure was brought by farmers, farm groups, and others to provide public assistance in order to alleviate these problems, at the very time that the high cost public programs of the recent past were experiencing dissatisfaction and scrutiny by non-farm groups. These instabilities and stresses make the decade of the 1980s a crucial one in the structural evolution of the agricultural sector and the financial markets that serve this sector.

Origins of Agricultural Capital Markets

The historical development of agricultural capital markets in the U.S. is characterized by important innovations and refinements in financial institutions, financial instruments, and lending practices in order to make financial capital available to farmers on the most appropriate terms. In the late 1800s, American farmers relied heavily on country or frontier banks, local merchants, land mortgage companies, and life insurance companies to meet their financing needs. While often working well, these arrangements also had their disadvantages. Swings in credit availability were common, access to nonlocal sources of funds was limited and unstable, interest rates could be high and widely disparate among regions, and little public credit was available to cope with severe emergencies.

Since these "frontier" times, a number of significant developments in private markets and public policy have occurred to improve the workings of financial markets for agriculture in terms of the accessibility, reliability, versatility, and cost-effectiveness of credit and related financial services. Major policy developments have included: 1) authorizing and supervising the FCS, 2) developing special banking and commerce laws, and 3) offering direct credit services from the U.S. government through the Farmers Home Administration (FmHA) and the Commodity Credit Corporation (CCC). Other public programs, focusing on crop insurance and other types of insurance, have helped reduce farming risks and thus augmented credit availability for agriculture. Other finance-related developments have included the evolution of leasing arrangements for farm land and non-real estate assets, and periodic interest in and concern about the role of outside equity capital in agriculture. Finally, the evolution of commercial banking in the U.S. through the development of the Federal Reserve System, the emergence of a strong regulatory environment, and structural policies that have preserved the community bank system have contributed significantly to commercial banks' strong involvement in financing agriculture.

The current structure of agricultural capital markets is characterized by significant holdings of equity capital by farmers, considerable leasing of farm land, and strong reliance on borrowed funds by many farmers. While the types of credit institutions are relatively few, they have differed greatly in their organizational structures, geographic scope, operating characteristics, specialization in farm lending, legal and regulatory environment, and government affiliation. For non-real estate farm lending, most of the activities involve commercial banks and Production Credit Associations (PCAs) that, along with merchants and dealers and backup support by government credit, offer a fairly wide range of choices to farm borrowers. The range of choices has been narrower in long-term lending, where Federal Land Banks (FLBs) dominate and where considerable financing by individual sellers of farm land has occurred as well. Life insurance companies, commercial banks, FmHA, and miscellaneous lenders are also active in long-term lending, but on a smaller scale than FLBs and individuals.

Financial Structure and Credit Conditions in Agriculture

Evaluating the effects of financial markets and credit conditions on the structure and performance of the agricultural sector is a complex task. It involves social preferences about desired structural characteristics in farming and about the availability of abundant, low-cost supplies of food and fiber. It also involves preferences by farmers for reasonable levels of profitability, risk, and liquidity in their farming operations. These are not independent sets of preferences, and costly conflicts may arise from the tradeoffs between them.

First, consider the role of financial markets from the broad structural vantage point of an industry with a very large scale, corporate organization of production units versus a smaller scale, noncorporate organization of resource ownership and control. Here, the historic developments in agricultural capital markets have been important in helping to maintain a pluralistic, smaller scale, largely non-corporate organization of agricultural production units, that seemingly has been consistent with the public interest. Readily available credit has contributed over time to significant gains in agricultural productivity, to the mechanization and modernization of farming operations, and to more orderly marketing of farm commodities. Indeed, many of the changes occurring within the smaller scale, non-corporate structure of agriculture--fewer farms, that have grown larger in size, greater specialization, greater capital intensity, stronger market coordination--have been facilitated by readily available credit. Finally, the assured provision of credit through both good and bad economic times has been a valuable source of liquidity for farmers to cope with various sources of risk.

But, readily available credit has had its disadvantages as well in terms of the effects on the long-term profitability, risk, and liquidity of many farm businesses. This is especially true in the 1980s. Viewed in retrospect, the agricultural conditions of the 1970s and 1980s stand in sharp contrast to one another. Some observers would characterize these times as another episode of "boom and bust" in agriculture in which the buoyant growth of the 1970s was dearly paid for by the depressed farm income and financial stresses of the 1980s. To put these issues in perspective, a brief look at a longer history is helpful in understanding current stresses and their relationships to financial markets.[1]

As Melichar (1984) points out, it is helpful to return to the early 1900s and begin with the rapid growth in farm debt that occurred during the boom period of World War I and shortly thereafter. The ensuing hard times in the 1920s initiated a conservative posture toward credit use by farmers that lasted through the depression years even though credit and refinancing programs were vital in these times. Indeed, studies of farm bankruptcies in the pre-1930s showed a much stronger association with debt levels than was the case in more recent times when farm income levels had more prominent effects (Shepard and Collins). The debt repayment orientation continued throughout and beyond the World War II era until farmers began to expand indebtedness during the 1950s. The 1950s and 1960s were characterized by the mechanization, modernization, and growth in size of farm units (and reductions in farm numbers), and the related demands for credit accelerated. Real income from farm assets grew fairly steadily, as did land values, and experiences with debt financing proved favorable for many farmers.

Then came the booming, yet volatile, 1970s in which farm prices and incomes skyrocketed in 1972-73, and farmers' expectations were buoyed to the point where the prospects for continued prosperity seemed very bright. Food storages, depleted reserves, strong demand, especially from foreign markets, and other concerns about limits to economic growth were involved. Faced with the choice between a conservative versus a liberal attitude toward credit, many farmers favored the latter. Moreover, this seemed reasonable

> "...in light of two decades of rising real farm income and wealth, successful employment of financial leverage through debt financing, and greater public concern about future scarcity of natural resources" (Melichar, 1984 p. 5).

Thus farm debt grew dramatically in the second half of the 1970s, as did the values of land and other farm assets which served to collateralize the growth in debt. The financial markets clearly supported the growth in debt; however, these times also witnessed the sharp growth in government credit, especially through various emergency programs that responded to the tight credit conditions of rural banks and to the adverse effects of several natural disasters that, in retrospect, appeared less severe than those to follow.

By the 1980s, a highly fragile financial situation had developed in agriculture. High incomes had been capitalized into rapidly growing land values that in truth were based on the expectations (or hopes) of continued growth in returns to farm assets. Moreover, the capacity to meet debt servicing obligations on the rapid growth in debt was based on continuation of relatively moderate, stable interest rates on farm debt. But neither of these events materialized. Instead, farm income fell and interest rates surged from the activities taken to fight high inflation, including a shift in the implementation of monetary policy by the Federal Reserve System, and from the deregulation of interest rates for depository institutions.

Lower incomes and high interest rates stressed the debt servicing capacity of highly levered farms, weakened credit worthiness throughout the farm sector, broke the growth wave, and caused both the real and nominal values of farm assets (especially land) to decline. Growth in farm debt continued in 1981 and 1982, due primarily to repayment problems rather than the desire for higher levels of debt financing. These adverse conditions for farmers aided a continuing shift toward government lending. During 1982 and 1983, farm loan demand softened considerably due to the continuing income and interest rate problems, and to the reductions in operating expenses from acreage curtailment in the Payment-in-Kind (PIK) and other acreage reserve programs. Even then, however, slower repayments to many lenders and more loan renewals and extensions occurred.

Considerable attention in the 1980s focused on the extent of credit problems in agriculture, on farmers' and lenders' responses to these problems, and on the implications for public policy and future credit programs. In general, farm credit problems in the mid-1980s could be considered very serious, but still manageable, with considerable differences in financial circumstances among farms. Most surveys seem to indicate that the farmers in deep difficulty numbered between three and five percent of all farmers, depending on how the problem is defined. This is still two or three times the normal rate. The incidences of delinquencies, loans being liquidated, borrowers discontinued, workouts and bankruptcies all accelerated from past experiences.

The evidence also indicates that the distribution of credit problems varied widely among farmers according to their financial positions. Hardest hit were those highly leveraged producers who had borrowed substantial funds during the growth period of the 1970s, and then had to borrow additional funds to cope with the adversities of the 1980s. In general, these farmers are relatively few in total number, but make a substantial contribution to farm production and commercial activity in agriculture (see Chapters 2 and 3). Moreover, the credit problems are more concentrated for lending institutions, since their lending activities only encounter those farmers who borrow money.

Clearly, these credit problems and the concentration

of debt in the hands of a significant portion of commercial scale farmers constitute an important issue in the agricultural capital markets in the 1980s. How well suited are the credit markets to deal with the near-term financial stresses in agriculture, as well as with the evolving forces in financial markets? And, over a longer term, how will the reshaping of these financial markets, due to regulatory and performance factors, affect agricultural financing in the future? These items are considered in the following sections.

Commercial Banks and Agricultural Financing

Commercial banks have long played a substantial role in providing financial services to the farm sector. However, it is difficult to generalize about this role due to the wide differences among banks in their structural characteristics, location, and proximity to agriculture. These differences also influence the way in which current stresses in agriculture and regulatory changes in financial markets are affecting these banks' performance as well as the evolving future patterns in agricultural banking.

The large money center banks located in New York, Chicago, San Francisco, Dallas, and other large cities generally finance large production units (cattle feedlots, broilers, etc.), agribusinesses, international trade in farm commodities, and also may participate in loans with other large banks. Regional banks located in medium-sized cities have also provided direct loans to large agricultural operations and agribusinesses, and have engaged in loan participations with smaller banks. Smaller, community-oriented unit banks located in rural areas have been the most heavily involved in farm lending. These banks operate in highly localized markets for funds acquisition and lending. They are especially vulnerable to local changes in farm income, which influence loan demands, loan payments, and deposit activities. Prior to interest rate deregulation, these banks also experienced periodic disintermediation problems that pressured fund availability for lending, and they relied on loan participations with correspondent banks to help meet large farm loan requests and growing loan demands.

One approach followed by the Board of Governors of the Federal Reserve System in identifying smaller banks heavily involved in farm lending is to define an "agricultural bank" as having less than $500 million in total assets and at which total farm loans account for 25 percent or more of total outstanding loans. Although the number of such banks is declining, they still totaled about 4,300 in 1984, accounting for 30 percent of all commercial banks, and holding one half of the farm loans in the banking system even though they hold only six percent of total banking resources (Melichar).

All these types of commercial banks have experienced the stresses of farmers and other economic sectors as well as the effects of deregulation of the U.S. financial markets. Key questions and issues involve the banking industry's responses to a changing, uncertain, dynamic financial environment, and how these changes will impact on agricultural finance.

Overview of Regulatory Changes

The major areas of regulatory change for commercial banks and other depository institutions in the U.S. have involved: 1) the decontrol of interest rate ceilings on deposits and loans, 2) structural controls on ownership forms and geographic scope, 3) the range of products and services that depository institutions may offer, and 4) the imposition of uniform federal reserve requirements on all types of depository institutions.[2] The major legislative changes accomplished by the mid-1980s were the Depository Institutions Deregulation and Monetary Control Act of 1980 and the Garn-St. Germain Depository Institutions Act of 1982. These acts focused heavily on changes in reserve requirements and on the decontrol of interest rates. Heavy emphasis also was given to aid for ailing thrift institutions by liberalizing their lending and investment activities, providing backup financing facilities, and allowing for their acquisition and mergers across state lines and even between thrifts as conditions warrant. Beyond these changes, additional legislation in the mid-1980s affecting bank structures and banking activities was also under Congressional consideration.

Interest Rate Deregulation

The deregulation of interest rate ceilings on deposits, called Regulation Q, was intended to alleviate periodic disintermediation problems, enhance competition in financial markets, and provide more equitable investment opportunities to small savers. Greater impacts were expected on smaller banks that had greater funding problems in the 1970s and less flexible balance sheet management than regional and money center banks. Indeed, the historic insulation of rural banks from national financial conditions and the heavy reliance on low cost demand deposits diminished considerably in the early 1980s. The levels and volatility of these banks' costs of funds increased, and essentially all of the funding sources for agricultural banks have become rate sensitive, and thus subject to frequent change. More market-oriented pricing of loans and services is occurring at the bank, department, and customer levels in response to these interest rate risks. The entire awareness factor among savers and borrowers, both farm and non-farm, about interest rates and pricing in financial markets has increased considerably.

Banks of all sizes and types are moving toward more effective management of assets and liabilities and more sophisticated pricing of funds acquisitions, loans and services. On the increase are the use of floating loan rates, risk assessment and pricing, matching maturities of loans and funding sources, cost accounting, loan documentation, and market-segmented pricing, although larger banks and holding company systems appear to have a greater capacity to use these methods. The traditional "customer relationship" in banking is changing, too--away from emphasis on deposits and more toward revenue generation with borrowers.

One logical consequence of a deregulated, more competitive financial market environment is downward pressure on banks' profit margins and levels of profitability. For agricultural banks, this appears to be the case, although by the mid-1980s the profitability responses appear mostly due to higher loss rates in lending and the higher costs of responding to borrowers' financial problems rather than to the direct effects of financial deregulation. The after-tax rates of return on equity capital for about 4,500 agricultural banks remained fairly strong through the 1970s averaging 14 percent per year, and then came off a peak of 16 percent in 1980 to trend steadily downward--to 15 percent in 1981, 14 percent in 1982, 11 percent in 1983, and 9 percent in 1984 (Melichar, 1985). Further declines could well occur for the mid-1980s as well. Of course, just like farmers, considerable variation in performance exists among these banks; some have been hit hard, bankruptcy rates have increased, and more agricultural banks are on the regulators' problem lists. But, other banks have maintained strong performance. Moreover, the sharp increases in commercial banks' volume and share of nonreal estate farm debt during 1983 and 1984 show a turnaround in the strength of their agricultural lending, at least compared to other farm lenders.

As indicated above, the dominant factor in the declining profitability of agricultural banks has been higher loan losses, part of which is on farm loans. The ratios of loan losses to total loans for these banks stayed in the 0.10 percent range throughout the 1970s and early 1980s, before increasing to 0.67 percent in 1982, 0.93 percent in 1983, and 1.26 percent in 1984. Moreover, the loss rate nationwide on farm loans for these banks was 2.2 percent in 1984, and 1.8 percent for all states except California. But part of these higher loss rates is a by-product of financial deregulation. After all, many banks responded to deregulation by passing higher, more volatile costs of funds on to borrowers through floating rates and other loan repricing methods. This in turn contributed to financial distress for many borrowers, which then returned to the agricultural banks and other lenders through higher lending risks, loan delinquencies, and greater loan losses. Moreover, the tendency of banks to respond to these credit problems by spreading the increased loan risks over other borrowers through higher risk premiums in loan rates has widened the incidence of credit problems among borrowers. Thus the profit position and lending capacity of many agricultural banks could deteriorate further in the future as lagged responses occur to farmers' stress positions and as some banks have problems adjusting to financial deregulation.

Bank Products and Services

A second area of deregulation involves changes in products and services that depository institutions may offer. Many banks have actively sought to engage in a broader range of activities than previously allowed in order to compete with new non-bank entrants to the financial services industry. These activities include real estate brokerage, securities underwriting, insurance, equity participations, and other non-lending activities. This area addresses the uniqueness of banking and whether banks should retain their traditional identity and deposit creating function in light of their important monetary role. These issues are important and may indirectly affect rural credit, especially if new products compete profit-wise with lending. However, these issues are likely less significant to agricultural finance than other regulatory areas.

Geographic Structure of Banking

The geographic structure of banking is another policy issue in the mid-1980s with important implications for agricultural finance. A long standing U.S. philosophy has been to grant individual states the sovereignty to determine branching and holding activity within their boundaries. The McFadden Act of 1927 prohibits interstate branching (except for a few systems operating when the Act was passed) and reserves intrastate branching policies to each state. The Douglas Amendment to the Bank Holding Company Act of 1956 prevents Bank Holding Companies from buying or establishing out-of-state subsidiaries unless authorized by the states involved. The result is a diverse set of state limitations on branching and holding companies. Thus, states may be designated as unit banking states, limited branching states, or statewide branching states. Moreover, states have differed in their authorizations for single-bank or multi-bank holding companies.

Considerable attention in the early 1980s focused on liberalizing the geographic restrictions on banking. However, except for thrift institutions, Congress had only begun to address these issues in the mid-1980s. Instead, the recent approach mostly involved letting individual states initiate geographic liberalization, using existing reciprocal authorities, as well as providing greater discretionary relaxation through regulators. This approach essentially allows the drift of market forces to work. Within this setting, many institutions have aggressively sought to expand as witnessed by the development of regional banking markets, the creation of non-banks (banks that do not simultaneously engage in commercial lending and deposit-taking), continued developments in electronic funds transfer, and the rapid expansion of multi-bank holding companies and mergers in states that eased restrictions on these activities. All of this has occurred within the traditional framework of correspondent banking and within past authorizations for loan production and Edge Act offices that allowed national and international lending markets at least for the largest banks and borrowers. It also is occurring within a less geographically restrictive framework for savings and loan associations, credit unions, and other non-bank financial institutions.

Views differ on how geographic liberalization would affect credit services to rural areas. Past studies found little support for believing that changes in bank structure would solve the farm lending problems of the 1960s and early 1970s (Board of Governors, 1975). But other views suggest that geographic liberalization could positively affect bank performance and enhance credit services for agriculture. These views are based on evidence (U.S. Treasury) that: 1) actual and potential banking competition is greater in states with more liberal branching; 2) more liberal branching does influence bank performance; 3) unit banks in statewide branching states have used more of their resources for loans than similar banks in unit banking states; and 4) average returns on total assets decline with increasing bank size, are highest in unit banking states and lowest in statewide branching and are higher for banks in non-metropolitan areas.

The future prospects for geographic liberalization in banking suggest a moderate pattern of deregulation that avoids national branching, as occurs in many other countries. Moderate deregulation should impact more on small institutions than on larger ones. Thus, a substantial decline in the number of banking entities in the U.S. could occur, perhaps by one-third by the mid-1990s. However, public pressures will likely continue to provide various protections for smaller, community banks that are so prominent in states that currently prohibit branch banking. These will be higher performing, community banks that are well managed, well capitalized, and strongly localized in their services. These banks will give much attention to the competitive pricing of products and services, to segmenting markets and perhaps specializing in activities like agricultural lending; to maintaining strong, personalized customer relationships where possible; and to establishing funding and service relationships with other banks and financial institutions. Some may become formally affiliated with multi-bank holding companies while others will remain independent.

In general, geographic liberalization should bring greater concentration in banking with more dispersed lending by banks over larger geographic markets. It should bring greater diversity in loan portfolios, greater risk bearing capacity, and thus downward pressure on risk premiums in interest rates, although risk factors from greater lending competition may be offsetting. It should enhance credit services for agriculture, although more along the lines of commercial loans for commercial scale borrowers and consumer lending for small, part-time farmers.

The diversity of bank lending to agriculture should continue, but with a swing toward greater financing from larger, more sophisticated banking systems. This will vary among borrowers with the larger, regional, or nationally oriented systems continuing to seek the business of larger farm units and agribusinesses. Continued participation among banks and banking systems will occur, whether through traditional correspondent banking arrangements or the more contemporary "franchising" of selected services. Bank mergers and consolidations, along with greater use of loan participations and non-local funding, will continue to shift credit controls and loan decisions toward non-local sources. However, the continued availability of experienced, well trained loan officers in rural areas should maintain the focus on local servicing of agricultural loans, while still achieving greater uniformity in loan documentation, risk assessment, and other lending practices. These developments should benefit agricultural borrowers, although continued development of their skills in financial management and understanding of financial markets will be important.

Non-local Sources of Funds

A long standing issue in agricultural banking has

involved the access of small independent banks to stable, cost effective farm related sources of funds that are independent of their local market conditions. As indicated above, these community banks rely heavily on their local markets for sources of loan funds and lending. Over time, their funding positions have experienced periodical stresses due to: 1) the long term growth in farmers' credit needs, 2) loan requests that exceed the banks' legal lending limit (limits for nationally chartered banks were increased by the 1982 Depository Institutions Act), 3) seasonal patterns in loans and deposits, 4) liquidity pressures on loans and deposits caused by changes in farm income conditions, and 5) the periodic need to reduce risks in loan portfolios and to restructure banks' balance sheets. Most of these conditions will continue in the future and, except for the geographic issues discussed above, financial deregulation should have little direct impact on smaller banks' access to nonlocal sources of funds.

The main sources of nonlocal funds for banks have included correspondent loan participations, discounting loans with Federal Intermediate Credit Banks (FICB), sales of government guaranteed farm loans and, to a lesser extent, lines of credit with other banks, seasonal Federal Reserve borrowing, federal funds purchases, and loan participations. These sources have worked well on occasion, but many agricultural bankers have sought more appropriate funding mechanisms for farm lending. Correspondent loan participations had the most use and have worked well most times. However, they also tend to be costly sources of funds for many agricultural banks and their borrowers. Moreover, the correspondent banks' willingness to participate, and familiarity with farm lending have varied over time. One major effect of interest rate deregulation, however, is the closer match between the interest rates from the rural and correspondent banks, as well as greater experience by farm borrowers with high, volatile rates. In addition, the Farm Credit Act Amendments of 1980 gave rural banks improved access to discounting privileges with FICBs by placing other financing institutions, including banks, in much the same position as Production Credit Associations in borrowing from FICBs, and more clearly establishing the conditions for banks to use this service. However, use of FICBs has not changed much since 1980, since most banks have experienced stronger deposit growth and softer loan demand. Finally, developing a more stable long term lending capability is especially important for those banks that wish to offer full service loan programs for their agricultural borrowers. Developing agent services for life insurance companies and using repricing methods to counter interest rate risks are some of the issues involved.

Over time, the leadership of agricultural banking has expressed interest in developing a new permanent mechanism for nonlocal funding by small banks heavily involved in farm lending. Most of the proposals and ideas have involved a multiple-bank approach operating through commercial channels to give permanent size advantages to agricultural banks, help preserve the attractive features of unit banking, and enable these banks to cope with the larger size and regional-national orientation of other farm lenders. In the early 1970s both the committee on rural banking problems of the Federal Reserve Board of Governors and an agricultural credit task force of the American Bankers Association evaluated and recommended several organizational approaches for implementing a new intermediary. In the late 1970s, two money center banks also attempted to develop funding mechanisms for smaller agricultural banks. Neither program had much use due to the prevailing financial market conditions and to apparent communication difficulties within the banking system. Further study and interest occurred through the late 1970s and increasing experience was accumulating with a few multi-bank agricultural credit corporations.

The most recent thrust in nonlocal funding for rural banks occurred through the formation of MASI, an agricultural credit corporation affiliated with MABSCO Agricultural Services Inc., originally called Mid-American Banking Service Company. In brief, MASI is a service agent through which participating banks in a number of states may discount farm loans with a particular funding source in the nationally financial markets. At present (1985), the funding source is the Rabobank of the Netherlands through its New York office. MABSCO is owned by the bankers' associations of 12 Midwestern states of the U.S. The MASI portion of its operations began on a pilot basis and became fully operational in 1983 with expansion occurring to other states and eventual plans to seek additional funding sources as well as offering long term loans along with short and intermediate term loans.

The financial innovation represented by MASI is a significant landmark in agricultural banking and finance. It offers a permanent, cost competitive funding source over a wide geographic base with strong uniformity in risk assessment, credit evaluation, and loan documentation. In the long term, it (or a similar organization) could offer participating smaller banks an effective funding source that helps them remain competitive as geographic liberalization of banking continues. It remains to be seen, however, whether many rural banks will make effective use of MASI. Since MASI began operations at a time when fund availability at agricultural banks was relatively strong, its initial growth has been moderate. Moreover, looking at past experiences, the only other significant attempt to develop a wide-scale nonlocal funding mechanism for rural banks involved the formation of FICBs in 1923. The subsequent failure of commercial banks to use FICBs eventually led to the development of Production Credit Associations as an important component of the Farm Credit System and user of FICB funds.

The Farm Credit System

Similar to other financial institutions, the Cooperative Farm Credit System (FCS) is also caught up in the swift and significant changes in regulation, competition, and financial stress affecting the U.S. financial system during the 1980s. While less directly affected than depository institutions, the key long term issues affecting the FCS basically involve the trade-offs between the needs by U.S. agriculture for a specialized, reliable, nationallyoriented credit system with special funding privileges in the financial markets versus the trend toward greater openness in the financial system with less emphasis on regulatory preferences in funding and on mandated specialization in lending and other activities for the institutions involved.

To more fully consider these issues, it is helpful to briefly review the development and evolution of the FCS found in Chapter 4. As of 1985, the major legislative authority of the FCS is the Farm Credit Act of

1971, as amended. This Act updated and consolidated much of the preceding legislation that had developed over time in a piecemeal fashion. Then, the Farm Credit Act Amendments of 1980 further updated and improved the operation of the system and marginally expanded its scope of activities. However, the passage of the 1980 Amendments occurred at a time when pressures were mounting for a significant move toward the deregulation of financial markets and, for perhaps the first time in its history, the FCS encountered significant resistance and debate about some of its legislative proposals. The issues raised in this debate were consistent with the general arguments about degrees of regulations in financial markets, and have continued to command attention through the decade of the 1980s. In the following sections the regulatory issues affecting FCS are considered in more detail, and the closely related issues of the system's performance in financing agriculture are discussed.

Regulatory Issues

Much of the debate on the Farm Credit Act Amendments of 1980 involved the viewpoints of the commercial banking industry and the concept of a "level playing field" regarding the regulatory environment for these types of financial institutions (Barry; Wilkinson). Included in the debate were differences among institutions in their access to financial markets (the Agency Status issue), geographic restrictions, tax obligations, legal reserve requirements, lending limits, stringency of regulation and supervision, and the range of financial services and borrowing clientele for these types of institutions. Many of these factors are difficult to evaluate in competitive terms, since the institutions involved are charged with serving different clientele and providing different financial services. Indeed, similar to other bank and nonbank comparisons, many of these differences are involved with the uniqueness of depository institutions, and whether the monetary role of banks should be carefully delineated and protected.

None of these issues was fully resolved during the debate on the 1980 Act. However, the legislation that was finally passed did reflect some responses to the concerns raised by commercial bankers and others. These responses mostly involved the access of commercial banks and other financing institutions to borrowing privileges from FICBs.

Following 1980, considerable attention in policy circles focused on the "agency status" of the securities that FCS sells in the financial markets. The Office of Management and Budget of the U.S. Government initiated a study of the agency status issue in 1981 with a view toward shifting credit to a completely private status where possible. The focus has been on government agency securities in general, not just the FCS (other agencies include the Federal Home Loan Bank, Federal Nation Mortgage Association, and the Student Loan Marketing Association). More recently, the Grace Commission also recommended changes in agency status of these institutions, specifically suggesting that a fee be levied on the agency privilege in order to equalize interest costs among financial institutions.

Agency status for the FCS is largely a vestige of early times when the Federal government was significantly involved in administering and funding the System's loan programs. As indicated in Chapter 4, however, the government capital has long since been repaid and the agency securities have no government guarantees. Nonetheless, the historical roots, federal charter and specified economic functions of the FCS institutions have likely perpetuated the belief that the government stands behind their securities.

Another more significant characteristic associated with agency status is the set of regulatory exemptions and preferences granted these securities when the legislation for the various agencies was enacted and as regulations developed for other types of financial institutions (Barry, 1984; Lins and Barry). These regulations have continued after the FCS and other agencies reverted to a largely private status. The actions required to remove agency status have never been clearly defined, but they would involve changes in most or all of these regulatory items.

Several studies have considered the possible effects of removing agency status (Lins and Barry). While the magnitude of the results may differ for an abrupt versus a phased removal, the general consensus is that loss of agency status would increase the interest cost of FCS securities to levels comparable to those of high grade corporate bonds or commercial paper. This increase would primarily reflect a higher risk classification for the securities, as well as responses to changes in the tax status of the securities (interest income from FCS securities is exempt from state and municipal taxation but not federal taxation), and the tax status of various FCS units (income from FLBs and FICBs is exempt from federal, state, and local taxation). The magnitude of increase in interest rates for farmers could be one half to one percent, or even more if the securities fell into a higher risk class. In addition, the volume of marketable securities could be adversely affected as well, although this depends on the efficiency and depth of this type of financial market. The volume of FCS bonds issued has ranked second to the U.S. Government for issuances of individual entities, and is much above the annual volume of the largest corporate issuers. With no precedence for private issuances of this size, many bond dealers, analysts, and others familiar with these markets are concerned that removal of agency status could reduce the volume significantly. If this occurs, less available credit and higher costs would create a number of significant effects on agricultural production, farm income, land values, rural communities, as well as on the general economy.

A contrary view, however, is that even without agency status, the financial markets are efficient and deep enough and the FCS securities have had favorable enough record that the entire funding needs of FCS could still be met, although at higher interest rates. Indeed, while agency status has been significant in maintaining access to financial markets, the strong credit history of FCS, its sound financial structure, and an efficient, well managed distribution system for securities also are important. Given the system's strength in financial personnel, and other resources in funds management, it seems unlikely that market access would be severely jeopardized, especially if loss of agency status did not occur precipitously.

As Lins and Barry point out, the agency status issue will be resolved by the political process that in the mid-1980s, favors continuation of agency privileges, especially in light of the financial stress affecting agriculture. But, given the substantial role of FCS in farm lending, the increased competition in financial markets, and the movements toward reduced government involvement, it seems likely that attempts to remove agency status will continue. More than just the access to funds is involved. One viewpoint is that if FCS is mandated to serve agriculture in all states and regions and through all phases of the economic cycle, then it needs a reliable funding source in support of its role as a reliable lender. Agency status is believed to contribute importantly to this funding reliability. If, however, agency status is removed, then wider authorizations in funding sources (perhaps including deposit taking) and asset allocations may be needed for FCS to continue this role. Other relevant considerations involve the stability and liquidity that agency markets add to the financial markets, the regulatory advantages and disadvantages of other types of lenders, and the implications for federal revenues and borrowing costs. Agency status actually may be a low cost to pay for the stability and high performance provided by this credit system for agriculture.

Thus, the agency status issue has important policy implications that affect the financial markets in general, the farm credit markets in particular, and especially the costs and availability of credit from FCS. In turn, these effects have important implications for the structure and performance of the agricultural sector. These issues associated with agency status are closely related to the performance issues of the FCS, which are considered next.

Performance Issues of FCS

In general, the FCS is considered to have achieved very high performance in providing credit and other financial services to the agricultural sector. The system is one of the largest, most efficiently operated, and effective farm credit institutions in the world. Loan funds have been available to farmers, ranchers, cooperatives, and other eligible borrowers in a timely fashion, for a variety of purposes, and in amounts, costs, and maturities that compare favorably with other sectors of the economy. For these reasons, the FCS has become the dominant lender in the farm credit markets, especially in long-term lending where FLBs are sometimes the only active institutional lender.

Financial Stress in Agriculture

Since 1980, the FCS, like other lenders, have been heavily buffeted by the financial problems of agriculture. Loan volumes have declined for some units, higher loss rates have occurred, some borrowers have been discontinued, personnel changes have occurred, and some associations have merged due to financial problems. These responses, of course, are logical to expect in stress times.

The actual losses experienced by FCS have been minor over time, but are on the increase in the 1980s. Loan losses by FLBs, for example, were $1.5 million in 1982, somewhat higher than in preceding years, but very small relative to FLBs' outstanding loan volume of $47.2 billion at year-end 1982. In 1983, FLB losses increased to about $9 million, and then to over $100 million in 1984; they are expected to increase further in subsequent years. The loss rates among PCAs have been higher than FLBs, more volatile, and increased sharply from an average of $20.3 million for 1977 through 1981 to $159.2 million in 1982 to $236.6 million in 1983, and to about $240 million in 1984. Still, the 1983 and 1984 figures were only 1.2 percent of PCAs' loans outstanding at year-end 1983. The Farm Credit Administration also reported in December 1984 that through the first 11 months of 1984, 39 PCAs were merged with neighboring associations to strengthen their capital position. In addition, several PCAs were placed into liquidation, and a couple of the district Farm Credit Banks have had significant problems. Because of these stress conditions, interest rates to other system borrowers have come under upward pressure in order to generate additional income and maintain a sound capital structure that sustains the system's high credit standing in the financial markets.

The FCS will likely experience more losses in the future and, similar to the case of agricultural banks, a delayed response of losses to farmers' stresses likely will continue, especially in long-term lending. The acquired farm properties held by the system will continue to increase and they will seek orderly ways to return these assets to the control of agricultural producers and other investors. Despite these pressures, however, the FCS has a number of features that enhance its financial strength and competitiveness during this era of deregulation and agricultural recession.

The major responses by the FCS units to farm financial stresses in the 1980s included the use of existing reserves to cover loan losses, building additional reserves, reducing system-wide financial leverage, and placing greater emphasis on loan quality, credit control, and customer workouts. The ratios of loss reserves to actual losses have typically been high for FCS units compared to other types of financial institutions. To illustrate, as of year-end 1983 the total (FLB, FICB, BC) allowance for loan losses of $655.4 million was 73.8 times the average net charge-offs for the previous five years, compared to a ratio of about 3.0 for large commercial institutions. Moreover, due in part to slower loan growth in 1982 and 1983, the system-wide debt-to-equity ratio has exhibited a clear downward trend from 10.83 in 1979 to 8.48 in 1983. For PCAs in particular, total loans outstanding actually declined during the 1982 to 1984 period.

These financial adjustments, as well as others, have helped the FCS to cope with borrowers under severe financial distress. In early 1983, for example, the Federal Farm Credit Board reaffirmed the FCS policy of forebearance, that involves staying with a borrower if a workout appears possible. Included among the loan servicing techniques were deferred payments, amortized loan extensions, carry-over loans, refinancing, and partial liquidation on a voluntary basis.

Other elements of competitive strength for the FCS have included its nation-wide loan diversity, various types of inter-bank arrangements for risk bearing, growing emphasis on risk management, and the strong standing of the system's securities in the nation's credit markets. Nonetheless, however, the financial pressures on the system from farm financial stress have been mounting. By the mid 1980s the FCS stood at crossroads in terms of its capacity to deal successfully with farm credit problems. Important policy issues involve the possible need for various forms of federal assistance, the role of federal regulation and supervision, and the relationships to other credit and non-credit assistance programs for agriculture (see chapter 7 for a discussion of the possible policy responses).

Long Term Performance

Over the longer term, the FCS is clearly taking action to perform effectively in a more competitive, deregulated, financial environment. One such action during the 1983 to 1985 period was a significant self-study, called Project 1995, of the System's future missions and directions. This comprehensive study covered all aspects of FCS operations (agricultural financing, financial markets, government affairs, information and communication, and personnel) in order to identify appropriate strategies and policies for the future. Other actions have in general reflected the emergence of FCS as a vigorous, strong commercial entity seeking to achieve high performance on behalf of its member borrowers. Among these actions have been a stronger emphasis at the bank and association levels on the development and marketing of new products and services; the continuing trend toward centralization of management and other functions; the formalization of government affairs activities through trade association arrangements; and a moderately paced expansion of international activities.

Some specific examples of recent developments within FCS include: 1) formalization in the 1971 Act of loss sharing and participation arrangements among the various banks and associations; 2) development in 1975 of a joint service organization, Farm Bank Services, to provide research, information, and purchasing services for the 37 banks; 3) full adoption in 1979 of consolidated securities for issuance in the financial markets; 4) evolution of the system's fiscal agency into the Federal Farm Credit Banks Funding Corporation as the mechanism for selling bonds and discount notes for all 37 banks; 5) the development of a farm credit leasing corporation; 6) the development in 1983 of the Farm Credit Council as the System's federated trade association; and 7) continued unification of management and organizational structures within the banks comprising each of the 12 Farm Credit districts, and further unification of management and geographic boundaries for PCAs and FLB associations within Farm Credit districts.

The centralization activities in general are intended to gain efficiencies in operation, compete more effectively in loan markets, present a more coordinated-coherent structure to financial market investors, provide for a more unified response to policy issues as needs arise, and enhance the Systems risk-bearing and liquidity-providing functions. At both the bank and association levels, the trend toward common management and administration of short, intermediate, and long-term lending programs should have profound effects on loan servicing and lending competition, and enhance the capacity to deal with financial stress situations. Despite the trend toward centralization, however, the FCS personnel still believe that some degree of local autonomy is desirable in giving the system the flexibility to adapt effectively to changes in credit conditions in local lending markets.

The development of the Farm Credit Council (FCC) in 1983 represented a significant effort to remove the need for the Farm Credit Administration (FCA) to periodically assume what was largely an "advocacy" role on behalf of the FCS and its interest in various policy developments affecting farm credit. Indeed, the 1971 Farm Credit Act had redefined the relationship between FCA and FCS to reflect the conversion of the Farm Credit institutions from partly government owned to privately owned institutions, vesting management authority in the institutions and the distinctly governmental functions of examination, supervision, and regulation in the FCA. But, no mechanism was in place to collect and channel the institution's views on various policy issues. Thus, the FCC was formed in 1983 to serve as a federated trade association for FCS in order to bring relevant policy issues before the FCS membership and management, and to provide a mechanism for channeling the system's viewpoints and interests to the Congress, state legislatures, the general public, and the media on matters of governmental affairs. This development placed FCS on comparable grounds with other financial institutions whose views on governmental affairs are channeled through trade association arrangements.

In its international activities, FCS has become increasingly responsive to the opening of world agricultural and financial markets. It is involved in international activities through various forms of technical cooperation (training programs, assistance with AID credit projects, professional organizations), the secondary placement of its securities in international financial markets, and international lending. In lending, the Farm Credit Act Amendments of 1980 gave BCs the authority to finance the international transactions of U.S. farmer cooperatives. These lending activities began on a controlled basis in the early 1980s to assure that responsible and prudent lending practices were followed. The initial focus was on having the central BC be the leader in establishing the international financial network for the BC system with most of the export transactions supported by the guarantees of the CCC under their international program. Following this the international financing activities are expected to expand in volume with greater dispersal among the district BCs.

Equity Capital Issues in Production Agriculture

The long term philosophy in U.S. agriculture has favored keeping the ownership and control of farm units in the hands of individual farm families. As a result the farm sector largely has retained a noncorporate, small scale structure, in which the claims on equity capital (totaling $745 billion in 1984) have mostly been held by farm operators and landlords. For these traditional farm units, the major sources of new equity capital are the retention of farm earnings in the business, capital gains especially on farm land, and gifts and inheritances from other family members. Despite these general characteristics of resource ownership and control, intergeneration transfers over time have dispersed the ownership of farm land so that off-farm landlords and other individuals have also become significant equity claimants. Moreover, some subsectors of agriculture have departed substantially from the small business, family orientation.

In this section, a brief perspective on the role and use of outside equity capital in production agriculture, is developed. Then some of the important managerial and policy issues are considered. Relatively little data are available about outside equity, and many impressions are based on judgement and experiences of various analysts (Moore; Scofield; Sisson; Penson and Duncan; Barry, et al.; Lee, et al.). But the issues can still be clearly identified.

Within the traditional structure of agriculture, farmers, and ranchers have often pooled resources to permit a larger, more efficient operation. Some of these arrangements involved leasing of land and nonreal estate assets. Others have occurred as joint ventures, partnerships, mergers, or family corporations. In many instances farm families have incorporated their businesses in order to reduce income tax and estate tax obligations and other costs associated with intergenerational transfers of farm assets. While these arrangements cause structural changes in the existing business units, the total equity position of the parties involved largely remains unaffected.

Outside equity capital has entered production agriculture through both informal and formal arrangements. Informal entry occurs when outside investors buy land, for example, directly from farmers or other land owners. The investor generally arranges for management, labor, production financing, etc. to occur through leasing contracts or direct hiring. Professional farm management services often are involved, especially when the investment is in farm land.

Outside equity entering through formal market arrangements is indicated by incorporation and sale of stock, limited partnerships, and various types of agency services. These arrangements have been common in the equity financing of large cattle feeding operations in the Southwest and Great Plains regions of the U.S., as well as in cattle breeding, citrus, and vineyard operations. More recent channels either used or contemplated for attracting outside equity have included life insurance companies, pension funds, foreign investors, and equity participations by some lenders.

Outside equity may come from many sources, including individuals, associations, trusts, nonfarm corporations, private investors, or others with investable funds. Some entrepreneurs have focused on improving the mechanisms for flows of equity capital in and around local communities and sub-regions of states, believing that investable funds indeed are available in these areas and that local people might be more willing to invest than nonlocal people.

Most outside investors prefer not to participate directly in the management of farm units. They generally have well defined financial goals and prefer limited liability, reasonable liquidity, effective financial reporting systems, and undifferentiated claims on equity shares rather than direct ownership of farm assets.

Clearly, many types of traditional farm businesses are not well suited to satisfy these investment criteria. Moreover, outside equity can be an unstable financing source if equity is later withdrawn due to changes in profit prospects, tax laws, or other factors. A good example is the cattle feedlot industry that experienced rapid expansion and locational shifts in the 1960s, financed in large part by inflows of outside equity capital, through limited partnership and other arrangements, that in turn served as the basis for substantial leveraging. These structural changes in cattle feeding encountered sharp impediments in the early 1970s when declining beef prices and higher feed costs of 1972 and 1973, as well as more stringent tax laws, caused withdrawals of capital from existing cattle partnerships and few new partnerships were formed. In turn, cow-calf producers experienced financial stress from financing the placement of their own calves in feedlots, rather than selling to custom feeders.

Many prospective outside investors seek tax shelters on high nonfarm incomes. In the past, tax shelters in agriculture have involved the availability of cash accounting, current deductions of some capital expenditures for developing orchards and ranches, potentials for converting ordinary income to capital gains, favorable estate tax considerations, and pass-through of income losses to the investor, as with limited partnerships or sub-chapter S corporations. However, changes in tax laws over time have modified and made less attractive the ability of investors to benefit from these tax provisions (see Chapter 8). Nonetheless, skilled entrepreneurs still manage to adjust the channels for attracting risk capital to agriculture.

In looking to the future, key issues involve the public attitude toward the investment of outside equity capital in production agriculture and the effects of such investments on the profitability, risk and other performance areas of farm businesses. Should the attraction of outside equity continue to be closely scrutinized and regulated to protect the structure of agriculture? Or, should equity attraction be more open? In the past, efforts to facilitate the flow of outside equity into agriculture have met considerable resistance, as in the case of investment bankers' attempts to develop mutual fund types of investments in farm land and agency services for investments in farm land by pension funds. But this social resistance may be changing.

These issues rest heavily on the various needs for equity capital and the effects on resource allocation, resource control, risk spreading, and market coordination for commodities and inputs. The needs for outside equity capital largely are based on the combined effects of providing additional financial capital to acquire high cost capital items and accelerate business growth, and farmers' greater flexibility in organizing the financial structure of their businesses and adjusting this structure as conditions warrant. The emphasis on these factors tends to change with financial conditions in agriculture. In the growth era of the 1960s and 1970s, outside equity was viewed by some observers as a way to facilitate the growth and modernization of farm businesses because the equity from retained business earnings could not keep pace with expansion pressures. Then, during the survival era of the 1980s outside equity was viewed as a possible means of relieving the stresses of high financial leverage and burdensome debt obligations for many farmers. In both these eras as well as in other times, farmers likely would benefit from having available a broader set of financing alternatives.

In looking to the future, none of these needs for equity capital will likely diminish. Instead, the needs may increase as the capital costs for establishing and operating farm businesses continue to be high, and as the business and financial risk positions of many farm businesses remain vulnerable. The availability of channels for attracting outside equity in reliable, cost effective ways may offer a relevant alternative for agricultural producers to consider along with the traditional sources of equity and debt capital. The more difficult task may involve matching the types of payoffs and other characteristics of agricultural investments with the types of financial incentives investors need to undertake these investments. Assuring satisfactory levels of current returns and capital gains, acceptable levels of risk, and reasonable liquidity may be especially challenging since these performance measures have varied considerably over time, and the past commitments of outside equity to agriculture have often had high instability.

Leasing of Agricultural Assets

As discussion in other chapters has shown, leasing is a primary method of financing the control of farm land by farm operators, and has growing consideration and use in financing the control of nonreal estate assets. For the latter, a financial lease is a form of intermediate term financing for machinery, breeding livestock, structures, etc., that has much the same effects on a farmer's cash flows as debt financing with intermediate term loans. The leasing effects may even be more favorable when the tax positions of the lessor and lessee are considered.

In general, leasing of any kind of asset is considered a form of financial leveraging since the lease creates a fixed financial obligation (rent) to be paid by the lessee for the use of the leased asset. Offsetting the added liability is an asset held in the form of the lease contract. The essential difference from outright ownership is that leases last for a stipulated period of time, after which the asset control reverts back to the owner. Within this framework, leases exhibit a variety of characteristics for pricing, maintenance and cost responsibilities, length, etc. In many cases tax considerations may strongly influence the leasing terms, while market forces, customs in the local area, and economic conditions of the leasing parties are also important.

The leasing of farm land occurs on either a cash rent or share rent basis, with the latter being especially important in crop production. Share leases allow the sharing of production risks between the farm operator and the landlord along with the sharing of management responsibilities and operating inputs. Hence, share leases are highly favored in risk management since they allow the rental obligation and earnings from crop production to move up and down together. Within this sharing, many unique arrangements and innovations have occurred to tailor the leasing arrangements to specific farming situations. Moreover, in recent years various combinations of cash rents and share rents have been considered to keep the risk sharing characteristic, but avoid the need to share management responsibility. This has been important for those absentee landlords who have little experience with or interest in farm business management, but still wish to have an agricultural investment.

A shortcoming of the farm land leasing market is the long standing custom for many leases to be short term (e.g. one year), and for a continued reliance in some cases on oral rather than written agreements. In the past, these arrangements worked well since the parties involved often were well acquainted and even family members. However, the dispersion of land ownership among diverse investors, the strong competition by farmers for leased acreage, and the growing needs for greater stability in resource control have all combined to encourage more formal, longer term leasing contracts and the further development of secondary leasing markets. This is especially true for commercial scale operators who lease multiple tracts of acreage from several landlords. Moreover, it is closely related to the issues considered in the previous section about attracting outside equity capital into production agriculture. A common practice is for the outside investor to contract for production, marketing, and other services through leasing arrangements. In this situation, more formal arrangements and longer terms are generally preferred.

Leasing arrangements for nonreal estate assets have included the traditional custom operations, such as planting, harvesting, and other fieldwork, short term operating leases, and longer term non-cancellable financial leases. Most of the interest has been on financial leases, even though their use in production agriculture is much less than in other economic sectors. Generally, leasing in agriculture involves machinery, equipment, some buildings, breeding livestock, and dairy cows. Most manufacturers of farm machinery and equipment offer leasing programs to farmers through dealerships, although these programs receive only moderate use. In addition, leasing services are provided by some financial institutions, such as commercial banks and Production Credit Associations, and by independent leasing companies and individual firms or persons offering specialized services.

Except for price variations of the asset, the lessee acquires essentially all of the benefits, risks, and costs of ownership without having to make the usual investment of equity capital. In many ways, the financial lease is comparable to a credit purchase financed by an intermediate loan that provides complete financing, although prepayments of rent or security deposits are similar to down payments in a credit purchase.

From the farmer's standpoint, the low use of leasing is partly because financial leases have been a higher cost form of financing than ownership. A major factor favoring ownership is the owner's use of investment tax credit, rapid depreciation, and interest deductions in generating higher tax credits earlier in the investment period compared to leasing. This has made the present values of cash outlays for ownership less than leasing. An offsetting factor to the high cost of leasing from leasing companies is the growing interest by financial institutions in offering to farmers (and others) intermediate term financing under financial leases at lower cost than under loans. The financial institutions can do this through their tax savings from owning the leased assets. Since their corporate tax rates generally are higher than farmers, they may benefit more from the investment tax credit and accelerated depreciation associated with ownership. In turn, they can pass along part of this tax savings to lessees through lower rental payments.

In the future, the financial leasing of nonreal estate farm assets should mature into a routine type of financing that is favored by some producers and not others. Its use will vary with the tax and other considerations that affect the financial performance of the leasing parties. In turn this will improve the flexibility farmers have in financial planning and offer another financing alternative. Secondary markets for financial leases also may develop to add liquidity to the lease contract through transfers among lessees. Moreover, guarantee and insurance programs for farm loans offered by government agencies could be extended to cover the leasing arrangement. This would reflect the comparability of leasing and borrowing as similar forms of financing, and provide for their uniform treatment in public programs.

Summary

This chapter has sought to identify and appraise the key issues of the 1980s that are affecting the private sector components of the financial markets for agriculture. Primary attention was given to the role of commercial banking, the Cooperative Farm Credit System, and the markets for equity capital and leasing in agricultural production. The coverage has not been exhaustive, however. Little attention was given to the role of other farm lenders such as life insurance companies, merchants and dealers, individuals, and finance companies. Because farm lending is not these other lenders' major activity, their responses to credit conditions in agriculture mostly center on pricing loans to cover lending costs and on adjusting the magnitude of their lending up or down to reflect the swings between favorable and unfavorable prospects for farm profits and farmers' credit worthiness. Merchants and dealers have rather unique problems as well, since their credit programs are heavily involved with merchandising activities; moreover, in stress times they sometimes become last resort lenders to borrowers who are discontinued by other lenders.

Clearly, the regulatory and competitive changes of U.S. financial markets, along with financial stress conditions in agriculture are having profound effects on the cost and availability of financial capital for agriculture. Virtually all of the major farm lenders have experienced significant changes. The decontrol of interest rates has directly affected commercial banks, along with continuing changes in the geographic scope of banking and the range of products and services banks can offer. These changes have much more closely linked agricultural banking and farm borrowers to conditions in national and international financial markets. The Cooperative Farm Credit System has been significantly affected by the financial stresses of agriculture of the 1980s, and has experienced much debate about the agency status of its securities and its role as a specialized lending institution for agriculture. In turn the pressures on financial institutions arise in part from swings in the financial conditions of farm borrowers that are themselves influenced by financial market conditions.

The general swing toward deregulated competitive forces in financial markets means that the costs of borrowing for farmers will remain higher and more volatile than in pre-1980 times. The continued geographic liberalization of banking and the emergence of more complex financial systems will induce competitive efforts by lenders to segment the financial markets and will shift an increasing portion of credit control and loan authority toward sub-regional and regional money centers. This will continue to fragment and dichotomize the farm credit market, especially for commercial banks, so that commercial scale farm borrowers are treated as part of a lender's commercial lending program and smaller, part-time farmers are considered in consumer lending programs. This may have significant implications for the future structure of agriculture, and heighten the need for the commercial farmers of the future to nurture their credit worthiness and build their skills in financial management as they compete for credit services. Also important are the implications for public credit programs and credit policies to be considered in the next chapter.

FOOTNOTES

[1]The following discussion in this section draws heavily on the article by E.O. Melichar "A Financial Perspective on Agriculture" Federal Reserve Bulletin, January 1984, pages 1-13.

[2]Numerous other provisions were also included in the 1980 and 1982 legislation. These are covered in various review articles (e.g. Federal Reserve Bank of Chicago, 1981, 1983). It is also useful to note that "depository institutions," as defined in the 1980 Act, include commercial banks, savings and loan associations, mutual savings banks, and credit unions.

REFERENCES

Barry, P.J. "Prospective Trends in Farm Credit and Fund Availability." Future Sources of Loanable Funds for Agricultural Banks, Federal Reserve Bank of Kansas City, 1981.

Barry, P.J. Impacts of Financial Stress and Regulatory Forces on Financial Markets for Agriculture, National Planning Association, Food and Agriculture Committee, Washington, D.C., 1984.

Barry, P.J., J.A. Hopkin, and C.B. Baker. Financial Management in Agriculture, 3rd edition, Interstate Publishers, Danville, IL, 1983.

Board of Governors of the Federal Reserve System, Improved Fund Availability at Rural Banks, Washington, D.C., 1975.

Federal Reserve Board of Chicago. "The Depository Institutions Deregulation and Monetary Control Act of 1980." Economic Perspectives, Jan/Feb, 1981.

Federal Reserve Bank of Chicago. "The Garn-St. Germain Depository Institutions Act of 1982." Economic Perspectives, March/April, 1983.

Lee, W.F., M.D. Boehlje, A.G. Nelson, and W. Murray. Agricultural Finance, 7th edition, Iowa State University Press, 1980.

Lins, D.A., and P.J. Barry. "Agency Status for the Cooperative Farm Credit System." American Journal of Agricultural Economics, 66(1984):601-606.

Melichar, E.O. "A Financial Perspective on Agriculture." Federal Reserve Bulletin, January, 1984.

Melichar, E.O. Agricultural Banking Conditions. Federal Review Board of Governors, Washington, D.C., March, 1985.

Moore, C.V. "External Equity Capital in Agricultural Production." Agricultural Finance Review, 39(1979):72-82.

Penson, J.B., and M. Duncan. "Farmers' Alternatives to Debt Financing." Agricultural Finance Review, 41(1981):83-91.

Scofield, W.H. "Nonfarm Equity Capital in Agriculture." Agricultural Finance Review, 33(1972):36-41.

Shepard, L. and R.O. Collins. "Why Do Farmers Fail? Farm Bankruptcies 1910-1978." American Journal of Agricultural Economics, 64(1982): 609-615.

U.S. Department of the Treasury. Geographic Restrictions on Banking in the United States, Washington, D.C., 1981.

Wilkinson, D.E. "The Farm Credit System: Another Source of Loanable Funds." Future Sources of Loanable Funds for Agricultural Banks, Federal Reserve Bank of Kansas City, 1981.

7

Farm Financial Policy

Peter J. Barry
and Michael D. Boehlje

Government policy has a significant impact on the performance and structure of the agricultural capital markets. One effect is through regulation of the private financial institutions. But a more direct effect is through public lending institutions. The purpose of this chapter is to discuss farm finance policy with particular emphasis on the public credit institutions. The discussion begins with a broad perspective on the justification and evolution of public credit programs. The growth in public sector lending to agriculture since 1970 then is reviewed and the causes for this growth presented. The discussion turns next to the financial stresses of the 1980s and some of the short term public policy responses being proposed. Finally, the long term policy issues affecting agricultural finance are considered.

Perspective on Public Credit Programs

In a free market society, the preference is for financial markets and institutions to develop under private auspices to the extent possible in order to provide the major financial services of channeling savings into investments and allocating risks among savers, financial institutions, and investors. To aid in this task, considerable regulations may be needed to assure effective market performance due to the intangible nature of financial services, and the need for confidence, trust, and stability among market participants. In addition, the private sector may experience significant imperfections and gaps in providing financial services due to extraordinary risks, geographic dispersion, insufficient credit worthiness, weak information systems, few market participants, and other factors that increase the costs of financial transactions. Under these conditions, the public sector may respond by creating new financial institutions or instruments, direct or guaranteed lending programs, or other forms of regulations for private institutions. In the case of agriculture, the federal and state governments have responded in various ways, including the direct lending and loan guarantee programs of FmHA, SBA and specialized state lending programs.

The public credit programs have long played an important role in achieving social objectives for U.S. agriculture as well as for other segments of rural America. These programs help channel funds to selected regions and areas of the U.S. and to various types of borrowers; they have helped maintain the smaller scale, pluralistic structure of the agricultural industry; they provide financing opportunities for beginning and limited resource farms with prospects for economic viability in the future; they provide valuable liquidity for emergency situations; they transfer wealth to the agricultural sector through various forms of subsidy; they provide financial support for building the infrastructure of rural communities; and, in the case of the Commodity Credit Corporation (CCC), they contribute valuable inventory financing to promote orderly marketing of farm commodities.

In addition, from a policymaker's standpoint, credit programs are a popular, politically expedient policy instrument. They are relatively easy and cost effective to administer, as long as program demands are not growing too fast; the administrative and risk bearing costs are difficult to measure and effectively hidden from taxpayers; they are highly visible to constituents; they can be quickly developed for responding to ad hoc crisis situations; and they do not directly influence commodity and resource markets, although the secondary effects on asset values, income, and risk can be significant. Moreover, credit programs give the impression of financial soundness since loan repayment is intended with interest.

All forms of public credit provide a subsidy to the ultimate borrowers since the public sector is taking actions to alter the flow of funds, the cost of funds, and/or the incidence of risk bearing. In a strict sense, the very existence of a government program is a form of subsidy relative to the market's performance without the program. In practice, however, subsidies are usually considered in terms of the following: 1) credit availability--is credit available that would not otherwise be available from commercial sources? 2) interest rates--are the government's rates below rates from alternative commercial lenders? 3) credit terms --are loan maturities, repayment patterns, and security requirements more favorable than those provided by commercial sources? 4) insurance costs--are the costs of insurance or guarantees for commercial lenders lower than the costs of comparable insurance from commercial sources? and 5) other services--does the government provide other services (loan supervision, management guidance, record systems, and so on) that either are not available from commercial sources or are available from the government at lower costs?

Greater emphasis has been placed in recent times on estimating the level of subsidies provided by all federal credit programs to their respective borrowing sectors. These estimates, as developed by the Office of Management and Budget, are intended to reflect the relative amounts of wealth transfer occurring each year through the public sector to the ultimate borrowers. The subsidy on direct loans is the additional payments that would have been required on the loans if they had been purely private. Since these additional payments occur over the life of the loan, they must be discounted to present values in order to represent the subsidy provided by the government loan. For 1983, the subsidies of FmHA programs were estimated as $410 million for farmer programs (14 percent of direct farmer loan obligations), $1,688 million for rural housing (57 percent of direct housing loan obligations), and $254 million for rural development loans (34 percent of direct development loan obligations). The total subsidy of $2.35 billion for the FmHA program is 28.1 percent of the grand subsidy total, which is considerably higher than FmHA's proportion (16.3 percent) of the total obligations made by federal credit programs in 1983. (For CCC programs, the subsidy was estimated to be $672 million, or 5 percent of direct CCC loan obligations).

The nature of these subsidies along with their

politically attractive features result in public credit programs having highly sensitive socioeconomic effects on various economic sectors, as well as on the financial markets. While positive effects are intended, this sensitivity can sometimes swing the balance toward negative effects. Public subsidies, for example, can alter the incentives for resources to flow to and from a sector, especially when excessive risks and other factors may warrant the dissolution of some firms; they can stimulate additional borrowing that pushes up production capacity and puts downward pressure on resource returns; they can add to borrowers perceptions of risk bearing and thus stimulate additional risk taking; they can quickly become capitalized into values of land and other fixed assets; they can sometimes run counter to the stabilization and income objectives of other public programs; and in financial markets, they can exert "crowding out" effects on borrowers in other sectors. Thus, these programs experience the continuing dilemma of balancing their positive effects against the unintended negative ones, especially when economic, social, and institutional conditions are changing quickly and significantly.

Historical Foundations and Evolution

The period from the early 1900s through the 1930s witnessed the basic shaping of the agricultural credit markets that exist today in the U.S., with major developments occurring along several lines. One set of developments involved the government facilitation of institutional reforms and innovations affecting the private credit markets, especially in commercial banking and the Cooperative Farm Credit System. For commercial banking, this included the creation of the Federal Reserve System in 1913, and a series of legislatvie enactments that clearly delineated the uniqueness of banking (the Glass-Steagall Act of 1933), allowed the diverse geographic structure of banking (the McFadden Act of 1927) in which both large and small banks could operate within state boundaries, and initiated controls over the pricing and risk environment of banking in the 1930s through ceilings on interest rates, portfolio requirements, lending limits, and deposit insurance. These developments stabilized the banking system and helped sustain the heavy involvement in many states of small rural banks in financing agriculture.

For the Cooperative Farm Credit System (FCS), the first 35 years of the 1900s saw the development of the Federal Land Banks starting in 1916 followed by the Federal Intermediate Credit Banks in 1923, and Production Credit Associations and Banks for Cooperatives in 1933. Thus, the basic framework of FCS was completed during the 1930s, and the long term goal was to move the system toward a private status in which ownership and control would rest with the financial institutions and ultimate agricultural borrowers, funding would occur from the private financial markets, and government control would be limited to the regulatory and supervisory functions that other types of federally chartered financial institutions experienced. These goals have largely been attained since the 1930s and the FCS now operates with a mission, as stated in the 1971 Farm Credit Act, to "...accomplish the objective of improving the income and well being of American farmers and ranchers by furnishing sound, adequate, and constructive credit and closely related services...necessary for efficient farm operations." And, "...a permanent system of credit for agriculture which will be responsive to the credit needs of all types of agricultural producers having a basis for credit..." (emphasis added).

These early developments in the 20th Century set the broad parameters for providing loan funds from commercial sources to credit worthy agricultural borrowers throughout the U.S. Indeed, in the 1980s the success of these early developments is indicated by the combined share of total farm debt held by commercial banks and the FCS reaching about 55 percent, with the remaining debt from commercial sources scattered over life insurance companies, merchants and dealers, and individuals.

Another major area of development in credit policy involved the financing needs of agricultural borrowers who could not obtain credit from commercial sources. The characteristics of these borrowers have changed over time, but have generally included impoverished, destitute farm families, young farmers entering the sector, small yet potentially viable farms, and larger farms experiencing significant distress due to natural disasters and economic emergencies. These types of farmers do not have access to commercial credit; thus, they are considered candidates for public credit programs. In turn, the federal government has responded with a long series of legislative and agency developments that included the creation of the Farmers Home Administration in 1946.

Although special credit programs and other aids to low income farmers in certain emergency situations had been available since before World War I, the groundwork for institutionalizing federal credit programs for agriculture occurred during the depression years of the 1930s. The very titles (rehabilitation, resettlement, security, subsistence) of the agencies and legislation showed the needs of the period and the evolution of the programs as economic conditions changed. The Federal Emergency Relief Administration was authorized in 1933 with the Rural Rehabilitation Division to focus on farmer problems of those times. This Division was transferred to the Resettlement Adminstration in 1935 that had the responsibility of resettling low income farm families and providing them with financial and technical assistance. Rural Rehabilitation loans with appropriate supervision were intended to serve the destitute and low income farm families who could not obtain credit assistance from other lenders. The Resettlement Administration was transferred to the Department of Agriculture in 1937 and later that year became the Farm Security Administration, with responsibility for continuing previous programs as well as for providing long-term farm ownership loans, as authorized in the Bankhead-Jones Farm Tenant Act of 1937, and for administering the Water Facilities Act that provided loans for farm water systems.

The Farm Security Administration continued its activities through the remainder of the 1930s and during the World War II period, although changes in economic conditions were stimulating the need for new and improved credit programs for post-war agriculture. Under the Farmers Home Administration Act of 1946 almost all of the direct lending to low income farmers that had been carried out for several years under a variety of programs and agencies, was placed under the newly formed FmHA. The new agency was authorized to assume functions performed up to that time by the Farm Security Adminstration and by the Emergency Crop and Feed Loan Division of the Farm Credit Administration. These functions included: to make operating and production loans to farmers and ranchers; to finance the purchase, improvement, or enlargement of family size farms; to insure mortgages made by private lenders for purposes similar to those of the Farm Ownership Loan Program; and to make water facility and disaster loans (Halcrow).

Besides direct financing of farmers through insured loans and (later) guarantees of farm loans made by commercial lenders, FmHA also acquired expanding authorities for making emergency loans. Two concepts have emerged in emergency lending. One involves emergencies attributed to disaster conditions of major proportions affecting selected areas or regions, and usually attributed to natural causes (drought, hurricanes, tornadoes, floods, fires, and weather-induced cases of insect infestations and diseases). The disaster concept was introduced in the Disaster Farm Loan Act of 1949, and has been a part of FmHA programs since then.

The other emergency concept is based on severe stresses imposed by economic and credit conditions. This concept surfaced in the 1970s first with passage of the Emergency Livestock Credit Act of 1974 (to guarantee loans made by commercial lenders to financially distressed livestock and poultry producers--terminated in 1979) and then the Agricultural Credit Act of 1978. The 1978 Act introduced the concept of an economic emergency, defined as a condition resulting from a general tightening of agricultural credit, or an unfavorable relationship between production costs and prices received for agricultural commodities, which has resulted in widespread need among farmers for temporary credit. The Economic Emergency concept was intended to be temporary but was extended several times. Moreover, substantial debt still remained outstanding in the mid-1980s.

Beginning in 1949, FmHA received legislative authorization for an enlarged scope of financing activities that continued to increase in the following three decades. These authorizations basically represented an extension of the agency's credit programs to the nonfarm activities of rural residents and rural communities. Housing loans for farmers were first authorized in 1949 and later extended to nonfarm residents. Housing programs for senior citizens, interest-supplement, rental assistance, and farm labor housing are some examples of activities that broadened the program and contributed to the growth in housing activity (Herr and LaDue). Water facility financing was extended to nonfarm rural customers and then to rural communities. Financing of waste disposal systems was added as well. In the 1960s, FmHA also undertook the financing of rural development programs. This included economic opportunity loans to low income rural people for farm and nonfarm enterprises and aiding communities to attract new industry by providing community facility loans. Following the Rural Development Act of 1972, FmHA's involvement in rural development expanded even more. This Act authorized FmHA to guarantee loans made by commercial lenders for farming, housing and rural business and industry, and greatly expanded the loan limits and availability of water and waste disposal loans (Meekhof). FmHA could also make loans for rural community facilities such as fire departments, hospitals, nursing homes, and public recreation facilities. In some cases, grants could be made for water, waste disposal, and certain other programs, including the improvement of rural industrial sites.

A second public sector lender that provides limited service to agriculture is the Small Business Administration (SBA). The SBA is a permanent, independent government agency created by Congress in 1953 to encourage, assist, and protect the interests of small businesses, in part by providing loans. Funds for the SBA are provided by annual appropriations made by Congress. Legislation passed in 1976 (P.L.94-305, approved June 4) amended various provisions of the Small Business Act and authorized the SBA to make loans to small businesses engaged in farming and related activities. The legislation did not create any new SBA loan programs, but merely included farms among the industries eligible for SBA assistance.

The lending objectives of the SBA are to stimulate small business in deprived areas, promote minority enterprise opportunity, and promote small business contributions to economic growth. These objective duplicate or are similar to some of the obejectives of the FmHA, so the SBA and FmHA entered into a memorandum of understanding (published in the Federal Register, Vol, 41, No. 199, October 12, 1976), which set forth the guidelines under which the agencies would operate. A primary guideline was that the SBA will encourage potential applicants who have been or are FmHA borrowers to continue with that agency and will advise others who meet FmHA eligibility requirements to contact that agency. In other words, the two agencies agreed to avoid duplication. As shown in a later section of this chapter, these arrangements to achieve coordination and avoid duplication encountered significant problems and had limited success until further arrangements occurred in 1980.

The Commodity Credit Corporation (CCC) also provides significant loan funds to farmers, but CCC loans are made primarily to implement federal price and income support policy rather than to explicitly provide credit to farmers. Nonetheless, CCC loan funds are a significant source of financing for farmers and affect the demand for funds from other public and private lenders.

In general, then, the missions and scope of public sector credit have changed considerably over time. In farm lending the emphasis shifted away from the heavy welfare orientation of the 1930s, including resettlement, rehabilitation, and subsistence, toward financing the production, investment, and resource development activities of farms that lacked access to commercial credit but with time, work, supervision, and credit could become viable commercial units in the future. The programs now are intended to provide a safety net to the farm sector; they help to buffer emergency situations and facilitate longer term resource adjustments involving young farmers, limited resource farms, and others with prospects for economic viability. In addition, the scope of the programs has been broadened to provide various types of credit for rural nonfarm residents and rural communities that have lacked equitable access to such credit from commercial sources.

Growth in Public Credit: Issues and Causes

A basic tenet of public sector lending has been to focus on those prospective agricultural borrowers who are unable to obtain credit from commercial lenders due to emergency situations or weaknesses in fundamental credit factors that can eventually be overcome. In the strict sense, then, it is the commercial lenders that determine the population of borrowers who will seek federal assistance by virtue of these lenders' acceptance of some borrowers and rejection of others. The significance of this point lies in the fact that a complete appraisal of the applicants for public credit depends not only on the credit characteristics of the applicant, as appraised by the applicable lenders, but also on the characteristics of the lenders themselves in light of the current conditions in agriculture and in the financial markets. Furthermore, the very existence of the public credit program may alter the behavioral and other characteristics of borrowers and commercial lenders, and thus warrant consideration in the appraisal process.

To show these relationships, let us suppose that all potential agricultural borrowers could be ordered on a scale from most credit-worthy to least credit-worthy, as shown in Figure 7.1. This ordering would be based on the results of a comprehensive credit analysis of each borrower which considers commonly accepted credit factors such as personal characteristics, management ability, loan purpose, income and repayment expectations, financial position and progress, and collateral. In practice, agricultural lenders do conduct individualized credit analysis, based on these factors, although they rely heavily on generalized rules of thumb in categorizing borrowing into acceptable and unacceptable

FIGURE 7.1: CREDIT WORTHINESS SCALE

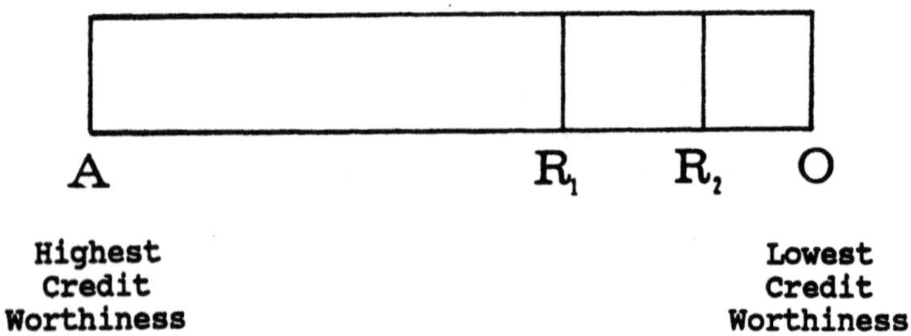

AR_1 = Commercially Acceptable

R_1O = Commercially Unacceptable

R_1R_2 = Last Resort Acceptable

R_2O = Last Resort Unacceptable

groups, and perhaps further categorizing acceptable borrowers for purposes of loan pricing, monitoring, and control.

Using the scale shown in Figure 7.1, four categories of borrowers can be identified. Set AR_1 includes borrowers that are considered acceptable to commercial lenders, while Set R_10 identifies commercially unacceptable borrowers. The members of R_10 then become candidates for public assistance, although within this group, only part, R_1R_2 are considered acceptable to the public lender, while the remainder, R_20 are unacceptable to any lender. Presumably, the public lender chooses its acceptable set, R_1R_2, based on the criterion that all of its members have the performance potential to eventually graduate into the commercially acceptable category, AR_1, while the members of the remaining set, R_20, have no viable prospects for graduation.

This credit evaluation process has uncertain and dynamic elements associated with it. Due to shortages of financial information, inadequate credit appraisals, excessive conservatism and so on, commercial lenders may make two types of errors: they may accept some unacceptable borrowers and reject some acceptable ones. Similarly, the public sector lenders can make the same errors in their acceptance and rejection of borrowers. The entire scale can shift locations, often quickly and significantly, relative to the stipulated boundaries as changes occur over time in agricultural markets or in individual farm situations. Moreover, the rejection line, for commercial lenders, R_1, can vary with events and forces in financial markets and individual lending institutions. Finally, the rejection line for public lenders, R_2, can vary with changes in social preferences, public budgets, lending regulations, and so on.

The public lender's role as a lender of last resort means that commercial lenders and perhaps commercially acceptable borrowers as well may view the public lender as a form of loan insurance. That is, significant deterioration in the conditions of less credit-worthy borrowers may make them eligible for the back-up support of public credit. The public credit program then may experience the same effects of moral hazard and adverse selection that weaken formal insurance programs. Moral hazard here means that the very presence of public credit may lead lenders, borrowers, or both to take actions that increase the likelihood of the program's use. These actions may include rapid business expansion and excessive borrowing, inadequate development of risk management by borrowers, low use of other policy instruments for responding to risk, and a higher incidence of loan rejections by some commercial lenders. The moral hazard problem is reduced by structuring last resort programs to distinguish among credit problems associated with 1) poor management and excessive risk taking, 2) uncontrollable events, and 3) particular types of borrowers for whom public credit is considered consistent with social objectives. Making these distinctions is often difficult at best in normal times, and especially troublesome under crisis conditions when the demands for public credit are growing rapidly.

Adverse selection essentially means that the public program draws participants whose risk position is higher than the average of those participating in the farm credit market. When this condition occurs, the costs of conducting the program are much higher than for the typical commercial lender who may also provide credit reserves for his borrowers. Adverse selection is, of course, an inherent characteristic of public credit programs, given their role as a lender of last resort. By the very nature of the program, public credit borrowers will have the highest risks and least creditworthiness among all active borrowers in the market.

Within this framework of credit concepts, dynamics, and uncertainty, the public sector must choose a level and form of participation in the credit markets that presumably is consistent with its stated goal as a lender of last resort for agriculture, and consistent with public preference for various structural characteristics for the sector. This of course, is a challenging task, especially in light of the relative ease with which public credit can be used as a liquidity response to crisis situations.

A major issue in farm credit markets has been the sharp growth in the level and market share of FmHA loan volume from mid-1970s to the mid-1980s, especially in nonreal estate loans. In evaluating the causes of greater public credit, it is helpful to utilize the concepts of credit evaluation, credit worthiness, and accept-reject considerations established in the preceding section. This framework suggests a number of working hypotheses that can be evaluated in light of recent changes in the financial conditions of agriculture.

The hypotheses include the following:

1. The growth in public credit is attributed to a weakening of fundamental credit factors in agriculture due to significant increases in business risks and financial risks. Included among the weakened agricultural credit factors are: a) lower income and debt repayment capacity, b) deterioration in loan collateral, c) excessive leverage, and d) weaknesses in farm management ability.
2. The growth in public credit is attributed to conditions in financial markets and institutions that caused wide swings in the cost, availability, and other terms of credit for agriculture, and thus altered the accept-reject boundaries for commercial lenders.
3. The growth in public credit is attributed to the slow development of risk bearing capacity in farm businesses and farm lending institutions.
4. The growth in public credit is attributed to the anticipation by farmers and commercial lenders that back-up support from public credit programs will be available, and these anticipations were reinforced over time as abundant public credit was forthcoming (the moral hazard issue).
5. The growth in public credit stressed the personnel and adminstrative capacity of FmHA, which was already dealing with greater authorizations for nonfarm lending programs, and likely increased the acceptance of some borrowers who a) could still qualify for commercial credit (those in set AR_1) and b) have no significant prospects for economic viability (those in set $R_2 0$).
6. The growth in public credit reflects the public sector's preference for and willingness to carry greater agricultural risks on concessionary terms.

Changing Financial Conditions in Agriculture

To evaluate the above hypotheses and assess their validity, it is helpful to consider the roots and evolution of the current financial conditions in agriculture, as summarized in Chapter 6. As that discussion shows, the last decade and a half has brought significant increases in the risk position of the agricultural sector. Business risks were the first to increase, due to the greater price volatility for major commodities, and then financial risks increased from the combined effects of high financial leverage and high, volatile interest rates. Moreover, the early 1970s also witnessed an attempt in the government's commodity policies to shift a large portion of the risk bearing function from the public sector to the private sector. Much of this risk rested initially with agricultural producers as opposed to input suppliers, product handlers, and retailers. In response to these higher risks, much emphasis in educational and other service activities was placed on building farmers' skills in risk management. However, progress has been limited in part because the business risks were very high, as witnessed by the large swings in farm income and the recessionary conditions of the early 1980s, and because of new financial risks from unprecedented interest rate and inflationary conditions. Thus, leveraged farmers have continued to rely heavily on the traditional financial responses to risk that include such practices as slowing down or deferring capital expenditures, drawing down liquid financial assets and even capital assets on occasion, and utilizing credit reserves with lenders to carry over loans, defer loan payments, refinance high debt loads, or otherwise provide liquidity. In general, these actions tended to create higher debt loads and have likely contributed to the increases in loan delinquencies, defaults, and foreclosures.

Responses of Public Credit Programs

In light of the changing financial conditions in agriculture, a strong relationship is clearly evident between the events that increased the sector's risk position and the increases in public sector lending activities, particularly FmHA. Moreover, the strong growth in public credit also appears attributable in varying degrees to the other causal factors hypothesized above: changes in financial market conditions, slow development of farmers' risk management methods, moral hazard, loan servicing pressures, and a liberal public attitude. These factors are reviewed here.

Beginning with the 1970s, one of the first significant responses by the public sector to the increased business risks in agriculture was the Agricultural Credit Act of 1974 that provided emergency loans from FmHA to aid financially distressed livestock and poultry producers following the sharp run-up in grain prices during 1972-73. This was followed by the Agricultural Credit Act of 1978 that expanded the emergency lending authorizations of FmHA and introduced the concept of an economic emergency. Included among the emergency conditions were tightness of credit and shortages of loan funds from farmers' commercial lenders. This credit provision was in response to the liquidity pressures and the credit stresses of smaller rural banks that had significantly increased their loan volumes during the buoyant times earlier in the decade. These banks were experiencing disintermediation pressures from other financial institutions and from money market mutual funds that slowed deposit growth as market interest rates rose above rate ceilings on bank deposits. These competitive pressures became even greater as inflation and interest rates rose and as the growth of unregulated money market mutual funds mushroomed. Thus, these adversities of rural banks contributed to the demand for public credit for farmers, as did the adverse effects of inflation-induced increases in production costs on the operating margins and cash flows of farm businesses throughout the country.

Another important development in public credit for agriculture in the late 1970s involved authorizations granted to the Small Business Administration (SBA) to also become involved in farm lending. SBA's involvement occurred fairly quickly and significantly and, while relatively short lived, had a strongly destabilizing effect on the farm credit markets over the long term. Some analysts would say that SBA's involvement in farm lending was a historical accident that originated primarily from the political process. There was a belief in the mid-1970s by members of the House Small Business Committee that FmHA operated under severe regulatory restrictions and had limited funding capacity for general farm loans, which in turn limited the financial assistance available for some needy agricultural producers. As a result, legislation was passed in 1976 that basically extended SBA loan authority to agricultural producers and provided for disaster loans at subsidized rates that could cover both actual losses and credit needed for ongoing production. Legislation in 1977 extended SBA's authority on farm disaster loans to also include crop losses. No credit elsewhere tests were necessary, as occurs with FmHA; the rate subsidies were substantial; the loan program operated through regional offices rather than through a county network like FmHA's; and the experience of SBA personnel with agricultural lending was essentially nil. Not surprisingly, the farmer demand for SBA loans was strong, and in some cases duplicated FmHA loans since two government agencies, with allegedly different purposes, were now active in farm lending. Beginning in 1980, regulations were established to better coordinate the agricultural lending activities of SBA and FmHA, and soon thereafter SBA's involvement in agriculture was terminated. Nonetheless, SBA's loan volume to farmers was considerable, reaching an estimated $3 billion of debt outstanding in 1981, with much of this debt still outstanding later in the 1980s.

The strong growth in the late 1970s of FmHA and SBA lending, especially through the emergency programs, with significant rate subsidies, clearly indicates the important role of public credit programs as a source of liquidity in these times. But this lending exhibited a number of different characteristics compared to previous public credit programs, and raised growing concerns about the types of farmers served, relationships with commercial lenders, the management and control of the programs, and the longer term effects on the financial position of farmers.

The characteristics of the farmers receiving public credit were clearly changing as a result of the emergency loan concept. A USDA study of FmHA borrowers in 1979 indicated that the traditional farm operating and farm ownership loans were directed principally to young farmers and those with small net worths and low incomes (Lee, et. al.). In contrast, large portions of funds in FmHA's economic emergency program went to farmers with relatively large net worths, high debt loads, and strong income potential (but low current incomes). Thus, the actual or potential availability of government loans on subsidized terms may have encouraged farmers to use or to become eligible for these programs. Moreover, while difficult to measure, the emergency concept also resulted in FmHA refinancing a considerable amount of problem loans for commercial banks, PCAs, and other lenders that served to alleviate, at least temporarily, deterioration in these institutions' financial conditions.

The strong growth in both FmHA and SBA loan programs was also heavily strapping these agencies' capacity to effectively evaluate, service, and control these loans. The rapid increases in loan volumes were not accompanied by commensurate increases in personnel or other agency resources, so that case loads of field staffs increased considerably. Moreover, the FmHA personnel also were responsible for other farm and nonfarm programs, and SBA personnel were responsible for the more familiar programs involving nonfarm small businesses. Delays in handling loan applications occurred and both agencies came under criticism for their lending practices. In the early 1980s, FmHA responded to these programs by adding field personnel, conducting more extensive training programs, and adopting more uniform, rigorous procedures for documenting loans and monitoring borrower progress.

Another problem with the FmHA emergency loans involved the procedures for declaring counties as victims of disasters or other emergency conditions that would make these areas eligible for federal assistance. The authority for such declarations was delegated in part to the FmHA state directors offices in order to permit a more timely lending response. However, experience soon indicated that locating this authority at the state level gave a strong incentive to the parties involved to declare a county a disaster area in order to attract and benefit from the use of low cost emergency loans from the federal government. A subsequent regulatory change shifted this declaration authority to the federal level in order to allocate emergency funds on a more objective basis.

While virtually impossible to measure, the moral hazard phenomenon was likely operating to some degree during the late 1970s. Both farmers and commercial lenders could easily perceive the strong growth in public credit programs as a broad, durable safety net for softening the effects of volatile financial and commodity markets. This would allow much of farmers' investment and marketing plans to proceed largely unabated. After all, the later 1970s still provided optimistic signals to farmers; farm income, although variable, was high and the appreciation in land values was substantial. Moreover, interest rate and other credit conditions in rural financial markets were still significantly insulated from the growing pressures in the national and international financial markets. Thus, the strong demand for public credit was consistent with the events of the times.

These strong demands, together with the shift to the emergency concept and the limited servicing capacities of both FmHA and SBA, brought these agencies under sharp criticism for excessively liberal lending in some cases, and unauthorized uses of loan funds by some farm borrowers. A 1981 report by the Inspector General's Office of the USDA contended that during 1979-80 FmHA extended nearly $2 billion in improper or questionable loans to farmers who had sufficient financial resources of their own and likely could have qualified for commercial credit. Some of the loans were alleged to be on overly favorable terms, while others may have been for unauthorized purposes such as to buy land or refinance nondelinquent debt. The report argued for better accountability in FmHA loan programs and more emphasis on loan processing and servicing, relative to loan volume. A report by the SBA Inspector General at about the same time (1981) cited similar and likely more severe problems with SBA disaster loans in which improper lending activity mostly involved loans for greater amounts than the actual damage incurred, misuse of loans by farmers, and issuance of loans by both SBA and FmHA to cover identical damage. Considerable recovery of these loans occurred and, as noted above for FmHA, personnel and servicing capabilities have been enhanced considerably since 1980. But these experiences clearly showed the hazards associated with liberal, rapidly growing public credit programs.

By 1980, the strong responses by public credit programs to the riskier agricultural environment of the 1970s had created a significant policy dilemma. Considerable interest arose for curtailing FmHA lending and restoring the agency's last resort role. The total scope of the agency's programs was to be narrowed and interest rate subsidies were reduced by basing most FmHA rates on the government's cost of funds. More emphasis was to be placed on the coordination of government credit programs with those of other lenders, including the graduation where possible of more credit worthy borrowers to commercial sources. These efforts to curtail FmHA lending were hampered, however, by the growing financial stresses in agriculture and by the political pressures involving farm credit.

Short Term Policy Options

Given the financial stresses of agriculture in the 1980s, a relevant question addresses the appropriate responses in financial policy. The agricultural sector is facing a new financial and economic environment. Adapting to that environment may require further government assistance to ensure an orderly, cost-effective process of adjustment in terms of financial and human losses. However, most analysts believe that in the intermediate-term agriculture must also adjust to excess production capacity and lower values for some agricultural resources, especially land. If this is true, then a public policy that impedes that adjustment would be very costly, and might result in long-term dependence on government assistance as well as continued government interference. These factors heighten the need to search for policy responses that are targeted to the problems of financial stress, are politically acceptable in an environment of fiscal restraint, and do not impede the long-term adjustments needed to maintain a productive, efficient, and financially healthy agriculture.

Public policy currently does contain a set of rules to resolve severe financial stress problems--the bankruptcy rules. Although bankruptcy may involve immediate liquidation of the assets and a discharge of the indebtedness of the farm (Chapter 7 of the Bankruptcy Act of 1978, Public Law No. 95-593, 92 Stat. 2549, 1978), it can also involve restructuring and rehabilitating the business under Chapters 11 or 13 of the bankruptcy law. A farmer who chooses Chapter 11, or possibly Chapter 13, bankruptcy proceedings becomes a "debtor in possession." In this situation, the farmer generally continues to manage and operate the farm, possibly under the surveillance of a creditor's committee (Looney, 1980). A trustee to manage the property is appointed only in rare cases, so the farmer can continue to operate the farm as long as he develops an acceptable debt reduction plan.

The bankruptcy rules specify how the private sector will share financial losses in case of a default by a debtor, but two fundamental issues remain. First, should the private sector--the creditor, the debtor, and other involved parties--absorb the full loss, or should the public sector share part of the loss through a government transfer payment program? Second, and probably more important, who in the private sector must absorb the majority of the loss? In many cases the financial losses will be transferred from the producer and his lending institution, who have been directly involved in the financial management and debt utilization decisions, to those less directly involved in those decisions. This latter group might include input supply firms and other unsecured creditors including many landlords. A fundamental question involves the fairness of this sharing of the financial losses due to debtor default.

A second rather blunt policy instrument for responding to financial stress in agriculture is a debt moratorium. This alternative would deny the use of foreclosure procedures against farmers who cannot make their principal and interest payments, cancel or defer interest and principal payments for a specific time, reduce part or all of the indebtedness, deny deficiency judgments for those who cannot make their payments, or use various combinations of these methods. In general, a moratorium would enable the financially pressed producer to gain temporary relief from the financial obligations associated with excessive debt. Most debt moratorium proposals cover a specific period of time, but they do not eliminate the eventual and definite commitment to repay indebtedness. Consequently, a key to the success of such proposals is the assumption that the financial condition of the firm and the industry will improve enough in the intervening period to repay the obligations. Debt moratoria have been used with limited success in previous periods of financial stress, specifically the 1930s, to relieve the financial pressure faced by farmers.

The major direct cost of a debt moratorium is the income foregone by the lenders during the moratorium period. In addition, serious concerns arise about the implications for the long-run performance of the financial markets. A debt moratorium could cause lending institutions to conclude that such a prospect has a higher probability in future periods of financial stress. Consequently, lenders who feel their earnings flow may be interrupted by future moratoria might seek compensation for the added financial risk through higher rates of interest. Furthermore, some borrowers would be denied credit even with adequate collateral if the moratorium negates the value of collateral in the credit extension decision. In essence, the use of debt moratoria would likely have chaotic results in the financial markets, higher interest rates for farmers, and reductions in credit availability.

Another policy response is the provision of loan guarantees from a federal or state agency to protect the lending institution from potential default by the borrower. A government loan guarantee would reduce the lender's risk, thus encouraging forbearance and loan restructuring. The guarantee might be conditional upon an approved plan of liquidation, or other more permanent solutions. Such a program is currently available from the FmHA; moreover additional funding could be made available for guaranteed loans to eliminate the need for unique legislation.

To be a permanent and effective solution, a loan guarantee program must be combined with other alternatives such as a systematic restructuring of assets or liabilities to reduce the debt obligation or increase a business' cash flow. Properly structured, a loan guarantee program may provide the time needed to implement other more permanent solutions and protect the resource markets from collapsing in the process. Without such a long-term solution, a loan guarantee program might be perceived as simply a "lender bailout." A variation of the loan guarantee program is to offer the lender a federal or state bond in exchange for the loan. Such a program transfers the responsibility for collection, as well as the debt obligation, to the government. It likely would be a higher cost program than the traditional loan guarantee programs from the FmHA.

Federally assisted debt restructuring is a proposal that has received wide-spread attention recently. In fact, most of the current legislative proposals are variations of the debt restructuring theme. The premise of this approach is that providing additional time to repay the principal would reduce annual obligations, thus enabling some farmers to cover these lower annual principal and interest payments. For those who still cannot meet their debt obligations, restructuring would provide additional time to rearrange the financial structure of their business including possible sales of assets. Most restructuring proposals involve the potential of a write-down of the debt obligation as a condition to obtain a federal or state guarantee. The key concept is to provide a government incentive for the private sector to implement workout plans. This will "buy time" so that these plans can be implemented rather than forcing the sale of assets and possibly collapsing the resource markets. For many financially-stressed producers, such a program may not be a permanent solution; rather, it is the first step in a long-run plan to adjust the asset and liability structure of the business so that the firm can survive.

As noted earlier, the increase in interest rates has been a severe problem for agriculture. Consequently, various policy proposals include interest rate buy-downs or other forms of subsidy which focus on reducing interest costs for farmers. Interest rate buy-downs can be implemented in many ways: as a direct government subsidy of interest rates for farmers, an increased tax write-off for farm interest payments, a public guarantee to reduce the risk faced by lenders and therefore allow them to charge lower interest rates to borrowers, and by using tax exempt revenue bonds to obtain lower cost funds for agriculture. Interest rate reductions would benefit farmers in the short-run, because interest has become a major cost of production, especially for highly leveraged farmers.

An alternative to interest rate buy-downs for agriculture would be monetary and fiscal policies that reduce the government deficit and thus the demands of the federal government on the capital markets. The resulting reductions in interest rates throughout the U.S. economy would benefit farmers in the same fashion as an interest rate buy-down. Furthermore, lower interest rates in general would have a significant impact on the demand for agricultural commodities. U.S. investments would be less attractive to foreign investors, thus reducing the demand for the dollar, which would result in lower exchange rates and increased export demand for agricultural commodities.

Debt restructuring may not be adequate for some producers; instead, asset restructuring including liquidation may be required to improve the chances of firm survival. By the mid-1980s, much of the asset restructuring has involved liquidations of real estate and other capital items. However, limitations on liquidity in rural communities frequently result in substantial liquidation losses. Other means of liquidation must be investigated and perhaps facilitated by public policy. For example, lending institutions might be encouraged to take the title of real property in lieu of debt obligations, and then lease this property back to the original debtor. By keeping the property off the market, this arrangement would reduce the chance of resource markets being depressed further. In addition, leasing the property back to the original operator would contribute to the efficient utilization of machinery, equipment and other resources. Through this process the lender can convert a nonperforming asset into an earning asset through rental payments. A government program of providing funds to the lender in the amount of the assets taken back in lieu of debt could provide the lender with the liquidity needed to conduct the program on a profitable basis. This program might require the lender to sell the acquired assets over a two or three-year period with the original debtor having a first option to buy. A similar program might be implemented by a state agency or a newly formed private sector firm funded through state or federal revenue bonds.

Again, one purpose of such a program is to stabilize resource values. A critical issue in the 1980s involves the public sector's role in regulating, monitoring or facilitating asset liquidations. Legitimate concerns have been expressed about some lenders who are encouraging cash sales of assets without considering the implications for producers or asset markets. Collateral values are declining in part because of forced sales of assets in thin markets where liquidity is lacking. More innovation is needed in the liquidation process, including possible assistance from the public sector.

Another alternative that might involve public policy is recapitalization. In many cases, the financial structure of the business could be significantly improved through an infusion of equity capital from outside the firm. The new equity could come from debt holders exchanging their obligations for equity positions in the firm, or from outside investors providing additional funds to reduce indebtedness. An equity infusion may appear difficult to achieve. Who would want to invest in a financially troubled firm? In some cases family members may be willing investors in order to continue a family business. An expected future inheritance of nonbusiness assets could be converted into current cash through sale to other family members. A nonfamily investor might be willing to contribute capital for a larger-than-proportionate share of the ownership of the firm. Some investors may be attracted by the tax shelter available from operating losses; under certain conditions, an operating loss is in reality an asset for investors in high tax brackets. Unused tax credits also may make the equity infusion more attractive for the investor.

The third source of an equity infusion is the lender. In some cases, the financial condition of the firm is so weak the lender will incur a significant loss if the note is called, foreclosure occurs, or the operator elects bankruptcy procedures. For example, a firm might have current cash flow problems because of high leverage and aggressive growth, but still have strong management and reasonable prospects for future earnings. In this case the lender may minimize losses or increase the chances for recovery by converting debt obligations into equity. This conversion reduces the current cash flow burden of excessive debt servicing and releases resources (both funds and management) for productive use to enhance current and future income.

The role of public policy in outside equity infusions or recapitalization may involve reassessing current legislation which discourages such arrangements. Many states have passed laws that restrict or prohibit outside equity investments in agriculture. These prohibitions or restrictions should be reassessed in the current financial stress environment. Alternatively, an entity with government-financed venture capital might be formed to assist with the necessary equity capital infusion into agriculture under terms that are more acceptable to both farmer and investor. This arrangement could be financed with state revenue bonds or federal funding. Such an institution might resemble the Agricultural Development

Banks used in many Third World countries. These banks use a combination of public and private sector funding that might be a significant institutional innovation in the U.S. capital markets.

Long Term Policy Issues

The long term policy issues relate to the public sector's involvement in agricultural finance. Included are the focus and direction of government lenders such as FmHA as well as the lending authorities, funding alternatives, and organizational structure of the Cooperative Farm Credit System.

Looking to the Future of FmHA

As was shown in the earlier sections of this chapter and in Chapter 4, the lending activities of FmHA have grown significantly since 1970, widened considerably in scope, and experienced substantial pressures from the recent financial stresses in agriculture. These conditions have focused considerable attention on the agency and placed its programs under close scrutiny. Fundamental questions have been raised about the proper role of public credit programs for agriculture, as well as operational questions about the appropriate forms of credit assistance to provide, and the practices followed, in program administration. Concerns have been expressed about the magnitude of the subsidies provided to rural areas, the resulting tax burden, the possibilities for shifting much of FmHA's loan volume to the private sector, the capacity of the private sector to absorb these loans, which borrowers should be served and not served, the length of this service, and the balance between federal and state programs (Barry, 1985a and 1985b).

As noted earlier, beginning in the late 1970s, significant efforts occurred to curtail FmHA lending and restore the Agency's last resort role. At that time, the curtailment of FmHA farm programs was based on several factors: 1) the high taxpayer costs of public credit and other farmer programs; 2) the rapid program growth of the late 1970s that increased administrative pressures and brought criticism of FmHA (and the SBA) for excessive lending in some cases, and unauthorized uses of loan funds by some borrowers; 3) stronger financial performance anticipated for the farm sector; 4) more moderate growth in farm debt; 5) expanded use of the revised federal crop insurance programs for disaster protection; 6) more effective use of loan guarantees in private credit programs; and 7) greater innovation by commercial lenders in countering business and financial risks in agriculture.

These curtailment efforts were thwarted by a number of factors. Especially important were the increasing financial stresses in agriculture that in turn caused deteriorating financial performance for many farm lenders. These conditions brought considerable pressures from farmers, farm groups, lenders, and others to provide public assistance to alleviate these problems. In addition, the revised federal crop insurance program has had relatively moderate use and little progress has occurred in shifting FmHA's financing from direct loans to guaranteed loans.

By the mid-1980s, a clear dichotomy had emerged in the policy arena affecting farm credit. There was the short term view held by many of a compelling need for additional credit assistance in some form (see the preceding section) to provide an orderly transition toward whatever type of agriculture was to emerge from these turbulent times. At the same time, however, the prevailing sentiments throughout the agricultural community and elsewhere point toward longer term efforts to reduce the size of FmHA programs, sharpen its focus, further refine its operating policies, shift more of its activities toward the private sector or to other public programs (including those at the state level), and in general reduce the reliance on credit to resolve financial problems in agriculture. Clearly, the mid-1980s is a watershed era for FmHA in particular and agricultural policies in general. How the agency and the agricultural industry will emerge from these times will set the tone of federal credit programs for agriculture for a long time to come. In looking to the future, we identify the following set of issues affecting FmHA programs that warrant further consideration and possible change.

Magnitude of FmHA programs The farm lending component of FmHA has experienced sharp growth in both loan volume and market shares since the mid-1970s. This growth indicates that the reliance on federal credit to cope with financial problems in agriculture has been at a very high level for a lengthy period of time. These conditions have two significantly adverse consequences. One is that reliance on additional credit to resolve the fundamental causes of farmers' financial problems simply will not be successful. The second is that excessive credit often granted on concessionary terms will exceed and erode the fundamental mission of credit as a liquidity providing mechanism to finance commercial activity in agriculture, and thus hamper the effectiveness of credit markets.

From the farmer's standpoint, additional credit is not a substitute for income. In fact, when income is low, additional credit may simply worsen the farmers' financial conditions. Instead, changes are needed in fundamental factors affecting net income and agricultural production. Over the longer term, these factors include strong commodity prices primarily coming from enhanced demand, both foreign and domestic, reasonable operating margins between revenues and costs, and a better matching of total production capacity to anticipated product utilization. Moreover, stronger emphasis is needed on methods for stabilizing farm income rather than relying so heavily on credit as a source of liquidity to cope with unstable farm income.

From the financial market standpoint, it is important to recognize that a credit market's primary function is to facilitate financial intermediation by adjusting the liquidity positions of savers and investors. Because these markets deal with intangible financial assets that rely on the borrowers' "promise to repay," high levels of confidence, trust, stability, and integrity are needed among the market participants for these markets to function effectively. Without these attributes, credit markets can quickly lose their effectiveness. When adversity occurs and near term repayment prospects are diminished, then the liquidity adjustments provided by the credit markets indeed may involve debt restructuring, loan deferrals, and other responses. However, these adjustments should occur without either significant income transfers or chronic use of these responses. Otherwise, some market participants will begin to count on these erosive uses of credit, with little burden placed on their own performance. In general, then, the larger is the intended subsidy for the afflicted parties, the less the assistance should be channeled through credit programs.

Form of credit assistance Prior to the mid-1980s virtually all of FmHA's farm lending occurred through direct loan programs rather than through guaranteed loans. This practice contrasts with the heavier reliance placed on loan guarantees by other federal credit agencies. The basic premise of a loan guarantee is that the resulting reduction in lending risks will encourage private sector lenders to make more credit available to higher risk borrowers at affordable interest rates without requiring the direct outlays of public funds. Of course, direct outlays may occur under cases of loan default. The guarantee may or may not be a part of a formal insurance program in which the lender, borrower, or both share in the program's cost, although cost sharing is a desirable long term objective.

Loan guarantees are believed to provide lower levels of subsidy than direct loans since the latter usually offer borrowers lower interest rates and longer loan maturities than are available from private sector lenders, even with a guarantee. Guarantees also provide less displacement of financial market resources and utilize the private lender's skills and resources for loan origination, analysis, documentation, monitoring, and control. The guarantees may also contribute to loan liquidity for the private lender if the guaranteed loans are more readily saleable in secondary market transactions.

Disadvantages of loan guarantees are the loss of direct control over lending activities by the public agency and reduced capacity for targeting loans to specific groups and in response to specific events. Thus, direct loans may be preferred as an initial response to emergency situations, or for specific types of borrowers with subsequent reversion of the direct credit to a guaranteed basis with private sector lenders. Finally, continued efforts should occur to refine the process of making guaranteed loans in order to minimize the time and handling requirements for the parties involved.

Types of farm loan programs In the past, FmHA has administered a number of farmer loan programs that have included farm ownership loans, farm operating loans, emergency disaster loans, economic emergency loans, and others. Each program has had its own purposes, funding authorities, loan limits, and other requirements. In order to facilitate lending activities in the future, and to keep pace with the evolution of financing programs by private sector lenders, the issue of consolidating FmHA's farmer programs into one financing program in which the FmHA personnel have the authority to structure a package of short, intermediate, and long term loans consistent with the needs of eligible borrowers might be considered. Consolidation would enhance financial management and control at the borrower level, provide for more complete supervision and monitoring of existing borrowers, and allow further tailoring of credit arrangements to the borrowers' needs. At the agency level, consolidation would allow greater flexibility in loan structuring, facilitate joint arrangements with other lenders, and allow more uniform supervision and counseling with various types of borrowers. Even the emergency loan programs could be integrated into the regular farm loan programs of FmHA in order to provide for a more uniform, complete approach by FmHA personnel to credit evaluation in light of changing conditions in local areas. This approach would still accommodate needed changes in the magnitude and terms of lending programs as selected geographic areas or types of farms become eligible for special assistance.

Users of FmHA services The general eligibility criterion for using FmHA services has essentially been to provide financing to farm operators who for various reasons lack access to private credit, but could eventually become economically viable and credit worthy. Among these borrowers have been younger farmers, limited resource farmers, and those who have suffered temporary setbacks due to natural disasters or unfavorable economic conditions. In recent times, however, the credit worthiness of many farmers has deteriorated to the point where private lenders are reluctant to continue financing them. These farmers have had sizeable operations, high financial leverage, and were hit by reduced income, rising interest rates, and falling land values soon after the expansion era of the 1970s. These farmers have sought public credit assistance as well, even though they differ considerably from the traditional concept of FmHA borrowers. The shift in the makeup of FmHA borrowers largely began in the mid 1970s as the result of the greater emphasis placed on emergency loans, especially those resulting from an economic emergency.

In the future, the general criterion cited above for FmHA lending (a last resort lender) likely will remain valid. However, greater emphasis could be placed on the borrowers prospective credit worthiness, economic viability, and eventual graduation to private sector financing. In addition, greater emphasis could be placed on those types of borrowers who are not eligible for FmHA services. To illustrate, the loan programs could be designed to clearly distinguish among these categories of borrowers: 1) borrowers who can either qualify for or continue to receive private sector financing; 2) borrowers who become ineligible for private sector financing due to events beyond their control, but have reasonable prospects for restoring credit worthiness and becoming economically viable in the future; 3) borrowers who experience financial adversity due to excessive risk taking and reckless, irresponsible management; and 4) borrowers with fundamental credit weaknesses and no prospects for future economic viability.

Category 2 borrowers likely are the most appropriate candidates for public assistance. They have the greatest potential for contributing to the sector's economic performance and thus to social well being. Category 4 borrowers may warrant termination from farming and consideration for nonagricultural assistance. They contribute little, if any, to economic activity in agriculture, although more general types of public assistance may cushion their adjustment from farming into other productive endeavors. Even if they remain unproductive, it is questionable whether agricultural programs should bear their costs, versus other more general welfare programs. Category 1 borrowers are best suited for private sector financing, while Category 3 borrowers are inappropriate candidates for any public assistance.

<u>Graduating FmHA borrowers to other lenders</u> While the ultimate aim of FmHA farm programs is to successfully graduate borrowers to private sector lending, the pace of graduation has been relatively slow and the process has tended to occur in a largely casual way. In the future, the approach toward implementing the graduation process could be upgraded and accelerated. One possibility is to place a maximum length on an individual borrower's participation in the farmer loan programs, after which

both the borrower and his indebtedness are obliged to graduate to private sector financing. Moreover, emphasis should be placed on graduation prior to the time limit to the fullest extent possible. To encourage graduation, definitive criteria and standards could be established to more clearly indicate the prospects for private sector financing. These criteria and standards should include commonly accepted credit factors and financial measures used by private lenders.

Periodic reviews of borrower progress should be strongly oriented toward graduation potential with greater consideration given to providing rewards and incentives for borrower progress. In FmHA's counseling and supervision activities with borrowers, greater emphasis should be placed on building the borrowers' skills in financial management in order to strenghthen their credit worthiness and enhance their prospects for permanent graduation to private sector lending.

Management and control guidelines for federal credit programs In 1984, the Office of Management and Budget substantially revised the guidelines to be followed by federal agencies in managing and controlling federal credit programs. These guidelines stipulate that existing and new credit programs will experience rigorous annual review of objectives, stronger justification, explicit statements of the borrowers' subsidies, pricing policies related to market interest rates and insurance premiums, a preference for loan guarantees, and private sector sharing of risk bearing costs to the extent possible.

None of these procedures directly curtail credit programs or limit lending capacity. However, these procedures should enable more effective controls on credit deliberations in the policy process. The subsidy measurements and annual reviews may moderate the rapid pace of change that some programs experienced in the past. Greater flexibility should occur for shifting credit to private markets where appropriate. Pricing relative to market conditions will enable program costs to more closely track these conditions and help in measuring subsidies. Finally, these procedures will help determine whether credit is the most appropriate form of assistance.

In general, these improved procedures for managing and controlling federal credit programs could foster a more conservative, purposeful approach to changes in public credit programs, and could enhance program effectiveness over time. These benefits appear to hold for the FmHA programs, as well as for other federal credit agencies. Thus, these control procedures will themselves benefit over the long term from an expeditious, efficient implementation by the respective agencies.

FmHA's housing and development programs Besides its farm programs, FmHA also received authorizations beginning in 1949 for an enlarged scope of financing activities. These authorizations basically extended FmHA's credit programs to the nonfarm activities of both rural residents and rural communities. Housing loans for farmers were first authorized in 1949 and later extended to nonfarm residents. Water facility financing was extended to nonfarm rural customers and then to rural communities with financing of waste disposal systems added as well. In the 1960s, FmHA also became involved in the financing components of rural development programs. This included the provision of economic opportunity loans to low income rural people for farm and nonfarm enterprises, and aiding communities by helping to attract new industry and by providing community facility loans. The latter included loans for such purposes as fire departments, hospitals, nursing homes, and public recreational facilities. By 1984, these nonfarm loan programs had grown to significant levels. The outstanding loan volumes were $27.4 billion for housing programs and $7.7 billion for rural development programs compared to $24.3 billion for farm loan programs.

The broadening of FmHA's scope over time to include housing and development loans represents a significant issue in evaluating the agency's future directions. Clearly, the housing and development loans have been important. This is indicated by their significant growth, and the close economic relationships among agriculture, rural communities, rural employment, and the investments in infrastructure needed to sustain the economic viability of rural areas.

However, agricultural, housing, and development loans still exhibit considerable diversity in loan evaluation, administration, and control that counters the effective dispersal and administration of these loans under the auspices of a single agency. Thus, further consideration is needed about the possible benefits of reallocating the nonfarm loan programs either to other federal agencies, to the auspices of individual states, or to some combination of these possibilities. A strong case could be made for separateness in policy deliberations, program development, and loan administration for the various types of credit programs affecting agricultural production, housing, and rural communities. This differentiated approach would likely provide for greater flexibility and precision in evaluating the magnitude and form of assistance needed by different borrowing groups, the preferred administration of the resulting programs, and the balance of responsibilities between federal and state levels.

Forbearance, credit responsibilities and rights. The current high delinquency rates for FmHA farm borrowers raise questions about the costs and benefits of forbearance and enforcing loan repayment commitments. Humanistic concerns suggest that a policy of forbearance is appropriate when severe financial stress makes it impossible for even the most efficient and financially astute borrowers to repay their loan obligations. But indiscriminate application of a forbearance policy may incur significant long-run social costs if current and prospective borrowers perceive that repayment performance is, in fact, optional. The high default rates in government credit schemes in developing countries clearly demonstrate the consequences of lax enforcement of repayment commitments. When access to credit is perceived as a right without repayment responsibilities, the borrower begins to regard the arrangement as a grant rather than a credit transaction. If this perception is reenforced through limited enforcement of repayment commitments, other borrowers will question their repayment responsibilities as well. Such an attitude typically is unacceptable to private sector lenders, so graduation from public to commercial lending institutions is less likely. The clear implication is that forbearance must be applied on a discriminatory basis; a blanket forbearance policy incurs a high risk of undermining the very concept of credit. Thus, future developments in FmHA policy should treat forbearance as an exception to be applied selectively, rather than as a common rule.

Supervised credit Much of the early focus in FmHA was on "supervised" credit programs that involved the provision of supervision and advisory services along with income, and their liquidity positions. The traditional commodity programs containing price supports, deficiency payments, and storage have focused on the level and stability of farm prices and incomes. In turn, crop insurance programs have been designed to respond to shortfalls in crop yields for individual farmers. The scope of these programs has been limited, however, in terms of the commodities and events being covered. As a result, credit programs have provided a last resort source of liquidity for selected types of borrowers and for coping with emergency situations. These credit programs essentially have filled the gaps left by the commodity and insurance programs. Moreover, the crisis to crisis nature of emergency credit make these programs vulnerable to heavy demands and political pressures for responding to the immediate problems of farmers and other affected groups.

These conditions suggest that a more comprehensive, coordinated long term approach for responding to income, stability, and liquidity problems in agriculture is warranted. In the process, the scope and magnitude of credit programs as sources of liquidity can be evaluated relative to the missions of other policy instruments in terms of their overall effectiveness in achieving social objectives for the agricultural sector. This long term assessment might indicate that a shift in the focus of agricultural policy toward greater use of insurance concepts in policy design is desirable. The insurance approach would enable farmers and others to bear more of the program costs, and could enable more comprehensive insurance protection to displace emergency credit, as well as replacing or modifying the need for other policy instruments that have selectively addressed price and production conditions. In turn, this would preserve and protect the integrity of credit markets in achieving their basic, liquidity providing missions, and more directly respond to the fundamental economic problems in agriculture.

The Farm Credit System

As developed in Chapter 6, two major policy issues affecting the Cooperative Farm Credit System involve its performance in responding to financial stresses in agriculture, and possible changes in the agency status of credit. For many inexperienced, beginning farmers, more assistance is needed than subsidized credit alone. In fact, studies have indicated that only beginning farmers with above average managerial capabilities have a reasonable probability of success (Thomas and Jenson). Part of the thrust of an institution serving beginning farmers should involve managerial assistance as well; without this focus the success rate of the institution will likely be significantly lower than desired. During the 1980s the supervised credit activities of FmHA have been reinstated through a commitment to recruiting "volunteers from knowledgeable, active and retired farmers and ranchers who will act as guidance counselors and provide direct, on-site farm or ranch management assistance" (Block). Having the capacity to continue such activities in the future is an important component of FmHA programs.

Credit Terms The interest rate, repayment schedule, and loan-to-value ratio are important aspects of implementing a public sector credit program. Historically, direct public loans have included an interest rate subsidy which reflected, in part, the "income supplement" dimension of these programs. More recently, interest rates on government loans have more closely reflected market rates to most borrowers, although rate subsidization still occurs in some cases. A key concern with the subsidized rate is the incentive to borrow more funds than would occur if market rates were charged. Furthermore, it is difficult to encourage public sector borrowers to move to private sector lenders when they can qualify for lower rate loans. It is also not clear how much subsidized interest rates benefit loan performance. A better procedure for reducing the cash flow and repayment pressures is to lengthen the term of the loan, thus reducing the annual principal payment, rather than lowering the interest rate. Deferred or variable repayment programs might also assist beginning farmers.

One hundred percent financing Much of the FmHA programs have focused on highly leveraged farming operations. In fact, the provisions of FmHA credit extension allow 100 percent financing under specified conditions. Even in the best of economic times, 100 percent financed operations will frequently encounter cash flow and financial problems; in periods of general economic stress, 100 percent financing is almost a guaranteed disaster. Furthermore, very high leverage positions present significant liquidity problems when funds are invested in farmland which historically has generated much of its total return as capital gains. Even with relatively low interest rates, land investments seldom are financially feasible with 100 percent financing. A fundamental issue in FmHA credit terms involves the realistic probabilities of success for highly leveraged borrowers who obtain 100 percent financing to acquire farm real estate.

State vs. Federal Programs

The recent interest in the development of credit programs at the state level for beginning farmers raises the issue of the role of the state versus the federal government in such programs. A primary goal in the creation of both the Federal Land Banks and the FmHA was to help young people move up the agricultural ladder from tenant to owner-operator. The pioneering work of the FmHA in developing low equity loans, cash flow lending, management supervision, and nonreal estate loans with terms of over one year were significant in giving entrants the needed financial tools. During the 1950s and 1960s the availability of the FmHA programs was often accepted as adequate public provision for beginning farmers. In this earlier period of agricultural adjustment policy debate was more often directed toward helping existing farmers leave or adapt their operations than toward helping young farmers.

The 1970s were marked by increasing criticism of federal policies that affect beginning farmers. The Farm Credit System was criticized as neglecting young farmers in favor of more secure loans to expanding established farmers (U.S. Senate). By this time the FmHA had grown to a giant agency which adminstered not only loans to small and low income farmers, but also credit for rural housing, industrial development, public facilities and economic emergencies. Less attention was available for beginning farmers. Thus more emphasis had shifted to the state level.

Research indicates that there is justification for developing state programs for beginning farmers. States can choose the type of approach most effective for their area, and they can choose the level of funding that matches the wishes of the voters who are demanding that program. Because support for beginning farmer programs seems to vary widely between areas of the country, state programs are likely to please more voters than a single nationwide program. However, federal level action has benefits in some cases. Credit programs may be most efficiently operated at the federal level because of economies of scale. Federal grants to state beginning farmer programs may also be needed because state programs create positive externalities.

However, the revenue bond funded direct loan approach chosen by most state legislatures for beginning farmer programs exhibits very high costs. The direct loan programs are attractive to the state legislatures because of low state cost, but the overall public cost is high and is paid for through reductions in tax revenue for the federal treasury. An analysis of various public financed, state-run beginning farmer programs indicates that a direct loan program for real estate control based on revenue bonds is about three times more costly than a guarantee program with a payment adjustment or a tax incentive program for real estate sale. The revenue bond method is about seven times more expensive than a tax incentive for leasing program (Lowenberg-Deboer and Boehlje).

Federal Credit Programs and Agricultural Policy

The long-term development of public policies affecting agricultural production and farmers could benefit from a more comprehensive, coordinated approach that considers the role of credit programs in dealing with agricultural risks, relative to other types of policy instruments. In the past, the various policy instruments designed to assist farmers have largely developed in autonomous, loosely coordinated ways that from the policymaker's perspective reflect their responses to the immediate problems and concerns of farm groups. Moreover, the expedience and popularity of credit programs have made them an easy tool to use, and likely prompted the significant build-up of FmHA credit in recent times.

To illustrate the possible relationships among the various instruments of agricultural policy, consider evaluating agricultural performance in terms of such financial criteria as the <u>level</u> and <u>stability</u> of farmers' its securities. Both of these issues warrant continuous monitoring and careful evaluation. Moreover, the two issues are closely related to one another since agency status helps the FCS to maintain reliable access to the financial markets, given its specialized mission of providing credit and related services to the highly cyclical agricultural industry.

In responding to financial stress in agriculture, most indications during the early 1980's were that unless farm losses became extremely heavy and widespread, the FCS should come through the stress times in reasonably good shape. Loan volume had declined for some units, higher loss rates were occurring, some borrowers were discontinued, the incidence of mergers among associations had increased, and new experiences were being gained with loss-sharing arrangements within and among the farm credit districts. Moreover, the system had taken several actions to strengthen its liquidity position and build its risk management capacities. Some of these actions involved continued internal restructuring of the systems capital positions and operations. Examples included the creation of central entities for funds management, risk assessment, and acquisitions of problem loans from district banks; the continuing consolidation of PCA's, including the move in several districts to a single district-wide PCA; and the shift to mandated intra-system loss sharing arrangements.

In general, the overall financial structure of the FCS remained relatively strong through the middle of the 1980's and the systems capacity to sell securities in the financial markets was not impaired. Nonetheless, policy makers, regulators, and others continued to maintain close surveillance of the systems performance due to its dominance in the farm credit markets and the important financial stability it provides to the financial markets in general. Finally, in September 1985, Donald Wilkinson, Governor of the Farm Credit Administration, with subsequent agreement by the leadership of the FCS, concluded and announced that substantial federal assistance could be needed in the next 18 to 24 months to keep the system solvent if farm financial conditions continued to deteriorate. The announcement was made in hopes that assistance mechanisms could be put into place on a contingency basis if and when they are needed. Various types of policy responses are possible. Some are short term responses focusing primarily on financial stress conditions; other responses have longer term consequences affecting the structure and performance of the FCS. Examples of these policy responses include the following: (1) Government guarantees of the system's farm loans and the farm credit bonds sold to investors; (2) the creation of a government entity to acquire non-performing loans and other assets; (3) a direct infusion of government capital into the FCS; (4) a buy-down of interest rates on FCS loans; (5) broadening the system's asset base through a wider range of loan authorizations; (6) allowing the merging of capital structures of the farm credit banks at the district level, and associations at the local level to provide greater system-wide uniformity; (7) authorizing additional sources of equity capital as with the sale of stock to investors in financial markets; and (8) authorizing changes in the number and makeup of the twelve farm credit districts in order to strengthen capital structures, build risk-carrying capabilities, and enhance management effectiveness.

Some of these same provisions affecting the flexibility of FCS are also involved in the agency status issue. Agency status basically consists of a set of regulatory exemptions and preferences affecting the System's sales of securities and the income tax obligations of various banks and associations. In looking to the future, the chances are high that policymakers will continue to face the issue of making fundamental changes in the agency status of FCS. This will especially be true when financial conditions in agriculture improve to the point where public credit assistance is no longer a major issue.

When this occurs, the consideration of the agency status issue may take the following forms: should the income tax exemption for Federal Intermediate Credit Banks and Federal Land Banks be terminated? Should a fee be levied on the Farm Credit System (and other agencies) for using the agency status privilege? Should FCS borrowers be required to cross comply with other federal credit programs, as in the case of commodity programs, soil conservation, and crop insurance? Should one or more of the following regulatory changes occur:

--should FCS securities be made ineligible for holding by federally supervised financial institutions, including nationally chartered banks, savings and loan associations, and credit unions?

--should FCS securities be subject to legal limits on holdings of individual issues for federally chartered depository institutions?

--should the exemption of interest income of FCS securities from state and federal taxation be terminated?

--should registration with the Securities and Exchange Commission be required for FCS securities?

--should FCS securities no longer be issuable and payable through the facilities of Federal Reserve Banks?

--should FCS securities be made ineligible as collateral for Federal Reserve Bank advances and discounts?

--should FCS securities be made ineligible for purchases and sales by Federal Reserve Banks as part of their open market operations?

--should nationally chartered banks no longer be able to underwrite FCS securities as part of the securities distribution process?

--should FCS securities no longer be eligible for the collateralization of public deposits such as tax and loan accounts?

The primary effects of changes in one or more of these factors would be to raise the costs (interest and noninterest) of funding the Farm Credit System (thus raising the cost of borrowing for farmers), and to create greater uncertainties about the supply of loan funds available to the System. In turn, if changes in agency status are considered, the Farm Credit System itself may seek to offset these changes through having a wider scope of activities in funding (offering deposit services, public equity offerings, and others), a broader customer base, and greater diversity in lending and other types of financial services for their customers. The broadening of scope would enable the System to gain further efficiencies in risk management to offset the higher risks of nonagency status, and to continue as a reliable lender to agriculture through all phases of the economic cycle. Moreover, this broadening of scope would be consistent with the trend toward greater openness in the financial system with less emphasis on regulatory preferences in funding and on mandated specialization in lending and other activities for the institution involved.

How the FCS copes with the turbulence of the 1980s will go a long way toward demonstrating their capacity to cope with changes in agency status. If the system comes through financial adversity in reasonably good shape in terms of its capital structure, reserves, and earnings, and if the system is able to effectively assist troubled borrowers through prudent policies of forbearance, debt restructuring, partial liquidation, and so on, then agency status likely will be regarded as less essential to the System's future performance. However, if the System itself experiences substantial adversities, perhaps to the point of needing public assistance, then agency status for its securities may remain an important attribute. The passage of time and further experience are essential in making these types of evaluations. In any case, the long- term trend, since the original creation of the various components of FCS, has been toward privatization of the System's ownership, operations, management, and financing, subject to the legislative authority of the Farm Credit Act of 1971 (as amended) and subject to the supervision and regulation by the Farm Credit Administration, an independent agency in the Executive Branch of the U.S. Government. Clearly, removal of agency status would be another and an essentially final step in the privatization process.

Equity Capital Issues

As indicated in Chapter 6, two of the major issues affecting the role of outside equity capital in agriculture involve the economics of risk bearing and greater flexibility for financing the future growth of farm businesses. In the stress conditions of the 1980s, the relationships between equity capital and risk bearing have received greater attention. Many farms in financial difficulty are well organized, of adequate size and use appropriate technology; however, their financial structure is characterized by excessive leverage. In a few cases, recapitalization may be possible and appropriate. This can be accomplished by adding equity from an outside source. One possible outside equity arrangement is a sale laseback of assets. In this case, debt would be repaid with the proceeds of the sale and business size would remain unchanged if the assets were leased back. Convertible debt instruments (such as convertible corporate bonds) or subordinated debentures may accomplish similar goals of giving the firm the financial flexibility to reorganize and improve the chances of survival.

Limited partnership arrangements also may enable highly leveraged firms to recapitalize and continue operation. Such an arrangement can be used to combine funds from several investors into a larger pool much like a mutual fund; the pooled funds then would be used to purchase financial interests in farming operations. Again, such arrangements could be structured to utilize tax shelter provisions, but would involve the pooling of funds from many investors for investment in a variety of farms to reduce the financial risks, compared to a single investor acquiring a single farm.

The "off farm" investor strategy for resolving financial stress is a controversial one. Many people are critical of the infusion of non-farm equity capital into agriculture, arguing that it will destroy the family farm structure and result in "outsiders" and "corporate" control of the production sector. In fact, some states have severely restricted non-resident alien investors as well as corporate ownership of farms in order to protect agriculture from "outside investors".

Several policy responses could be considered to encourage increased equity capital flow into agriculture. Restrictions on outside investment could be eliminated to open up the farm capital and asset markets to a broader group of participants. In addition, tax incentives could be established to encourage equity investments in agriculture. Such tax incentives might include the following: Increased flexibility in transferring and utilizing operating and capital losses; exemptions or credits on income earned on assets acquired from distressed farmers on a sale-leaseback basis; exempting capital gains of financially stressed farmers from the alternative minimum tax; and expanded tax sheltering potential.

Another policy option could involve the formation of a state or federally funded venture capital entity to share the private sector risk of equity investments in agriculture. This entity might play the role of an agricultural development bank. It might, for example, provide low cost funds or industrial revenue bonding for firms that buy assets from financially distressed farmers on a sale-leaseback basis.

An important and fundamental issue related to non-farm equity capital involves the property rights of owners (landlords) and users (tenants). Increased separation of ownership and control of real estate and other capital assets will have different implications depending upon the legal rights and institutional structure used in the farm rental markets. Changing the balance of property rights of tenants versus landlords, including the potential for longer term leases and compensation to the tenant for improvements made, may have a significant impact on the economic and social attractiveness of renting versus ownership of farm land. The institutional structure surrounding the rental of farm land is a significant function of property laws and public policy in general. If "reasonable" terms of trade are maintained between owners and users, the perceived negative social consequences of outside equity capital may be partially offset and substitution of non-farm equity for debt may improve the financial resiliency of the agricultural sector.

REFERENCES

Aronson, H.L. Agricultural Credit and the Small Business Administration, presented to seminar on Financing Agriculture in the 1980s, Agricultural Extension Service, University of Minnesota, Sept. 23-25, 1980.

Barry, P.J. and W.F. Lee "Financial Stress in Agriculture: Implications for Agricultural Lenders." American Journal of Agricultural Economics, Vol. 65, 1983.

Barry, P.J. Impacts of Financial Stress and Regulatory Forces on Financial Markets for Agriculture. National Planning Association, Food and Agriculture Committee, Washington, D.C. November, 1983.

Barry, P.J. "Needed Changes in the Farmers Home Administration Lending Programs" American Journal of Agricultural Economics Vol. 67, No. 2, May 1985.

Barry, P.J. "The Farmers Home Administration: Current Issues and Policy Directions" Looking Ahead, National Planning Association, Washington, D.C. Sept. 1985.

Block, John, "New Full-Time Family Farmer and Rancher Development Project," Secretary's Memorandum 1051, United States Department of Agriculture, October 21, 1981.

Boehlje, M.D. FmHA Farm Loan Programs: Issues, Objectives, and Consequences, paper presented to USDA workshop on Federal Credit Programs, Washington, D.C. November, 1983.

Brake, J.R. "Short Term Credit Policies for Dealing with Farm Financial Stress: Their Impacts on Structure and Adoption of New Technology," Office of Technology Assessment, U.S. Congress, Washington, D.C., 1985.

Congressional Budget Office, Loan Guarantees: Current Concerns and Alternatives for Control. U.S. Congress, Washington, D.C., August, 1978.

Congressional Budget Office, Conference on the Economics of Federal Credit Activity: Proceedings, U.S. Congress, Washington, D.C. April 10-11, 1980.

Halcrow, H.G. Food Policy for America. McGraw Hill Book Co., New York, 1977.

Harl, N.E."Restructuring Debt in Agriculture". Agri Finance, Vol. 26, No. 6, August 1984.

Harl, N.E. Debt Restructuring and Bankruptcy Proposals. Financial Stress in Agriculture Workshop, Federal Reserve Bank of Kansas City, Kansas City, Mo. Oct. 22, 1984.

Herr, W.M., and E. LaDue. "The Farmers Home Administration's Changing Role and Mission." Agricultural Finance Review, 41(July 1981): 58-72.

Hughes, D., S.C. Gabriel, R. Meekhof, M. Boehlje, D. Reinders, and G. Amols. Financing the Farm Sector in the 1980s; Aggregate Needs and the Roles of Public and Private Institutions. Staff Report, Economic Research Service U.S. Department of Agriculture, Washington, D.C., April, 1981.

Lee, J.E., S.C. Gabriel, and M.D. Boehlje. "Public Policy Toward Agricultural Credit." Future Sources of Loanable Funds for Agricultural Banks. Federal Reserve Bank of Kansas City, December 1980.

Lowenberg-DeBoer, J. and Michael Boehlje, "Evaluation of State Legislative Programs to Assist Beginning Farmers," Staff Paper No. 126, Department of Economics, Iowa State University, Ames, Iowa, January 1983.

Meekhof, Ronald. Federal Credit Programs for Agriculture, Economic Research Service, U.S. Department of Agriculture, Washington, D.C. 1984.

Melichar, E.O. "A Financial Perspective on Agriculture." Federal Reserve Bulletin, January 1984, pp. 1-13.

Office of Management and Budget, "Federal Credit Programs." Special Analyses Budget of the U.S. Government, Fiscal Year 1985. U.S. Government Printing Office, Washington, D.C.

Paarlberg, D. Private Sector Review, Loan Guarantees from the Federal Government, Review Team Approach. Financial Stress in Agriculture Workshop, Federal Reserve Bank of Kansas City, Kansas City, MO, Oct. 22, 1984.

Thomas, Kenneth H. and Harold R. Jenson, "Starting Farming in South Central Minnesota . . . Guidelines, Financial Rewards, Requirements," University of Minnesota, Agricultural Experiment Station Bulletin 499, St. Paul, 1969.

U.S. Senate, "Hearing on the Farm Credit Act of 1971," S. 1483 Committee on Agriculture and Forestry, 92nd Congress, 1st Session, 1971, p. 178.

U.S. Department of Agriculture. This is FmHA, Farmers Home Administration, August, 1980.

8

Tax Policy in Agriculture

Michael D. Boehlje

Taxes and tax management play a significant role in farmers' decisions about the production, marketing, and financial organization of their businesses. As a result, analysts generally can better explain or predict farmers' actions using after-tax rather than before-tax measures of net income. Furthermore, policymakers perceive that changes in tax rules will significantly alter saving and investment behavior as evidenced by the debate concerning tax reform in the 1980s, and the major changes in tax policy included in the Economic Recovery Tax Act of 1981. The purpose of this chapter is to evaluate the effects of tax policy on farm firm decision-making, aggregate investment behavior, and supply and prices of agricultural commodities.

Tax policy is included in this book's discussion of agricultural finance because of the integral linkage between the two. To illustrate, the tax treatment of interest expenses and the write-off procedures for purchases of capital assets affect investment and financing behavior. The differential taxation of ordinary income compared to capital gains along with the opportunities for tax sheltering in agriculture influence the types of assets purchased and thus the characteristics of the financing needed. Differential tax treatment of debt financing compared to lease financing affects the choice of an optimal financing option.

This discussion will first present some of the basic concepts of taxation. Then the institutional setting of the current tax law in the U.S. is reviewed, followed by a discussion of the concept of tax sheltering. The final section identifies some of the basic policy issues and discusses the relevant research in agricultural taxation.

Conceptual Framework

Taxes affect economic activity in the private sector in two fundamental ways:

1. they transfer resources from private individuals and firms to the government, reducing net income available to the private sector to spend or save while increasing the funds available for government spending or investment;
2. they change relative prices of different factors of production and different commodities (Institute for Contemporary Studies).

These private sector impacts are important in understanding the behavior of firms and individual investors, and in evaluating the macro impact of changes in tax law.

From a policy perspective, taxation has three key roles in society: (1) to impact the organization and efficiency of economy activity, (2) to redistribute income and wealth among members of society, and (3) to raise revenue to pay for government activities. These three roles provide the basis for evaluating changes in tax policy.

Efficiency and Resource Allocation

Efficiency in production and marketing has always been highly valued in agriculture. Most analysts indicate that consumers have benefited from improved agricultural productivity through lower cost, higher quality food. Tax policy affects the efficiency of agriculture through incentives or disincentives that the tax laws provide to: (1) acquire more productive technology, (2) substitute between capital and labor, (3) develop new technology, and (4) exploit economies of size.

The capital requirements in agriculture per worker and per dollar of sales are substantially higher than for most other industries. Much of the capital investment in agriculture embodies new technology which is in part responsible for the productivity increases in the sector. Tax provisions impact the cost of acquiring new capital inputs and thus, the optimal quantities of capital and labor to use. Tax provisions may also influence the long-run cost curves in farming and thus, economies of size. Finally, tax provisions can influence the rate of development and adoption of new technology. Recent examples include the tax credits and incentives provided for energy conservation and energy production from agricultural and other biomass products.

Equity

Equity has many dimensions; two of the most important to agriculture are opportunities to enter farming and the distribution of income and wealth. Tax provisions can impact the opportunities to enter farming by the incentives they provide for various individuals to buy and rent or operate real property and by their treatment of property transferred between family members from one generation to the next. Sizeable estate tax liabilities would reduce the ability of succeeding family members to continue farming (at least at the same scale as the previous generation) because of the necessity to liquidate farm assets to pay estate taxes. Lower tax liabilities can facilitate intra-family transfers, but may reduce the opportunities for entry by individuals whose parents or ancestors were not engaged in farming. Alternatively, tax laws might facilitate entry by providing incentives for sales of farm property from retiring farmers to beginning farmers.

A second dimension of equity involves an individual's "ability to pay." Horizontal equity requires that individuals with equivalent initial resource endowments pay the same taxes ("equal treatment for equals"). Vertical equity indicates how taxes vary as resource endowments or incomes increase. Policymakers' notions of vertical equity are reflected by the progressivity or regressivity of the tax system. A progressive tax system may be deemed desirable if a society believes that the rich should pay a larger share of their income as taxes.

While the market pricing system leads to an efficient allocation of resources, it may not be an equitable one. Welfare considerations and other national priorities may override the desire for economic efficiency. Society may deem it necessary to lessen disparities in income distributions while providing for economic stability and growth. Hence, the tax system's contribution to such policy goals must be considered along with its efficiency merits. Neoclassical economics indicates that welfare is improved if government expenditures are applied to equity concerns while tax revenues are collected in the most neutral manner possible.

Revenue Generation

Individuals with income from farming pay a small proportion of total federal income taxes. Any revenue increases from higher tax burdens on farmers could substitute for taxes paid by other businessmen or wage earners, and lower taxes for farmers would require increased taxes for other taxpayers unless government expenditures are reduced or larger government deficits are accepted.

Simplicity is also desirable in a tax system. The simpler is the system, the fewer are the opportunities for cheating, for legal tax avoidance, and for investment in projects strictly as tax shelters which are not profitable or economically sound. Simple tax codes make both administration and compliance easier and reduce recordkeeping requirements. The current tax system in the U.S. is not considered a simple one.

The Institutional Setting

Tax law is highly complex with a myriad of deductions, exemptions and credits. Furthermore, the law is frequently revised (witness the 1976 Tax Reform Act, the 1981 Economic Recovery Tax Act and similar laws passed annually since 1981) and new regulations, revenue rulings and court decisions continually update its application. Our focus in this section is not on the specific provisions of the law, but instead on the conceptual base for taxation of income, wealth, labor and property and the unique treatment of farmers. Empirical studies indicating the impacts of these laws on farmers and agriculture are discussed in a later section of this chapter.

The Income Tax

The U.S. income tax system raised almost 56 percent of federal revenues in fiscal year 1982, with $298 billion paid by individuals and another $49 billion by corporations. The income tax originated modestly with the Revenue Act of 1913 which imposed a progressively graduated tax on the income of individuals. The rates began at 1 percent on $20,000 of income and reached a maximum of 7 percent at $500,000 of income. Revenue needs of World War I pushed the top rate to more than 70 percent, but exemptions and allowances were so generous that most people paid no taxes. Since then, rates have been substantially reduced, and exemptions and allowances have been increased.

Besides the federal income tax, most states also impose an income tax. This tax is usually less burdensome than the federal tax, but in many states it is significant. In most cases, the state income tax has a structure similar to the federal tax.

<u>The Personal Income Tax</u> The individual federal income tax imposes progressive tax rates on an individual's net income each year. But if gross income and its related expenses can be reported in different tax years, the level of taxable income in each year can be distorted. Mismatching income and expenses in different tax years can be used to defer taxes and it can distort the application of progressive tax rates. Thus, many complex rules for reporting income have developed and most business taxpayers must use accrual accounting so as to properly match costs and receipts.

Special income tax rules for agriculture, however, permit farmers to mismatch income and costs, thereby enabling farmers to manage their tax liabilities. Such provisions include: (1) the use of cash accounting, (2) the immediate deduction of some capital expenses, and (3) the capital gains treatment for income from assets for which costs may have been deducted as a current expense. Cash accounting ignores the effects of changes in inventories during the year. Thus, even though the farmer has an increased inventory, he or she can deduct costs of inputs and control the tax year in which income is realized by the storage of crops and careful timing of sales.

Expenditures incurred in the development of certain farm assets, such as trees (other than citrus or almond trees), vines, and livestock herds used for draft, breeding, dairy, and sporting purposes are capital expenditures. However, farmers may deduct the full amount of such expenditures in the year when they are incurred. These expenditures can reduce or shelter ordinary income from other sources which would otherwise be taxed at regular rates. Then income from the sale of the assets is frequently treated as long-term capital gains with only 40 percent of the income currently subject to tax. This is one of several mechanisms for converting ordinary income to capital gains income.

Additional provisions can be used to reduce the tax liabilities of farmers and other businessmen. Expense method depreciation, made available by the Economic Recovery Tax Act of 1981, makes it possible to write-off the full purchase price of qualified capital assets in the year of purchase--up to certain limits. The accelerated cost recovery system (ACRS) enables the taxpayer to depreciate capital items over relatively short lives--3 years for automobiles, light trucks and breeding hogs, and five years for other breeding livestock, most farm machinery and specialized livestock and horticultural structures. Up to a limit, land clearing and soil conservation expenses can also be deducted as current expenditures. And investment tax credit in the amount of ten percent of the purchase price of five year ACRS assets and six percent for three year life assets provides a direct credit against the tax liability. The personal tax rates in 1985 (married and filing a joint return) range from 11% on taxable incomes between $3,400 and $5,500 to 50% on incomes in excess of $109,400.

The Corporate Income Tax The rates of taxation on corporate income, whether farm or nonfarm, differ from the rates on the income of individuals. Up to about $8,000 of taxable income, individual tax rates are less, but above this level, corporate income tax rates are less. In contrast, long-term capital gains of corporations frequently are taxed more severely than those of individuals. Under special tax provisions for corporations with a limited number of shareholders (the Subchapter S provisions), corporate income and partnership income are similarly taxed. It is reported by and taxed to the shareholders in their personal returns in the same form as earned by the corporation.

Net earnings of corporations that are paid as dividends to shareholders are again taxed when the shareholder receives the dividend. This individual income tax on the dividend is frequently called the "second tax" on corporate earnings. The second tax can be avoided if corporate earnings are retained in the corporation. A penalty tax is imposed on excessive accumulations of earnings that are not invested in business assets, but this tax can be avoided by using the accumulated earnings to expand the business. So the tax rules encourage farmers and agri-business persons to incorporate to save taxes. Moreover, once they have incorporated, the rules encourage expansion and growth of the business.

Operation of a business in corporate form can produce additional tax benefits. Fringe benefits such as group life insurance, housing, health insurance, and pension or profit-sharing plans can be provided to employees (including owner-managers). The costs of these benefits may be deducted by the corporation and usually are not taxed to the recipient employees even though they are also shareholders. In contrast, if these benefits were purchased by individuals operating an unincorporated business, their costs would be either nondeductible or deductible in a smaller amount.

Taxation of Cooperatives Farm input supply and product merchandising cooperatives receive special treatment under the U.S. tax code. In essence, this special treatment allows the income of qualified cooperatives to be subject to single taxation, even though it is distributed to the member-patrons. As noted earlier, in a corporate structure income is taxed as earned by the corporation and then the dividends distributed to the shareholders are taxed a second time. In a qualified cooperative, the distributions to the member-patrons are deductible to the cooperative, but they are taxable income to the member. The federal tax statutes do ensure that cooperative earnings are taxed at either the cooperative or member-patron level.

The historical justification for single taxation of cooperative income compared to the double taxation of corporations is that the cooperative is simply an extension of farmers' business operations. The cooperative provides goods and services to its members at cost, and any earnings are returned to partners in proportion to their patronage rather than to the number of shares they own in the cooperative.

Many cooperatives distribute a portion of their earnings to the member-patrons in a noncash form called patron retains. These patron retains enable the cooperative to accumulate funds for expansion and growth.

Regardless of the form of patronage distribution, the member must pay income tax on it. To minimize the potential problem of members having to pay more income tax on the patronage distribution than they receive in cash, cooperatives must distribute at least 20 percent of the distribution in a cash form in order to deduct the entire distribution from their taxable income.

Noncooperative business firms frequently attack the single tax treatment of cooperatives as unfair; they argue that cooperatives do not pay their "fair share" of taxes. Such arguments are substantiated by cooperatives that consistently pay the minimum amount of their distribution in cash, rarely revolve the patron retains to heir members, pay no interest on members' equities, are primarily growth and expansion oriented, and have moved into business activities only peripherally related to the business focus of their member-patrons. For those that indeed are extensions of the patrons' businesses, single taxation is likely justified.

The Estate Tax

The estate tax is imposed on a person's total wealth at death. Like the income tax, the estate tax is progressive; it is imposed upon the estate of a decedent and calculated as a percentage of it. Because of the graduated rate structure, the percentage of a decendent's estate which must be paid in estate tax increases with the size of the taxable estate. Until 1981 the maximum rate for large estates was 70 percent, but by 1985 the maximum rate was 50 percent, which is reached at the $2.5 million level of estate value.

Complementing the estate tax is a tax on gifts. It is imposed at the same rate and with many of the same deductions as apply to the estate tax. The gift tax is paid either in the year of the gift or the following year, and gifts bear tax at the full value of the property. The major purpose of the gift tax is to prevent people from avoiding the estate tax through making lifetime transfers. Many states also have inheritance or estate and gift taxes, but their rates are substantially lower than the federal rates. Federal gift taxes are subject to annual exclusions of $10,000 for gifts to any one individual. Moreover, the tax obligation on taxable gifts applies against the Federal estate tax credit, so that no tax is payable until the credit is exceeded.

As they grow older, property owners frequently want to give a part of their property to their families or other prospective heirs. They often find, however, that such gifts can be made only by giving away fractional interests in assets, which are difficult to arrange. In contrast, it is easy to make gifts of corporate stock. A gift of stock is, in effect, a transfer of part of the underlying assets held by the corporation without any change in title to them. Incorporation thus frequently facilitates transfers of partial interests in assets. In addition, the tax burden on gifts of corporate stock may be less than on a gift of the underlying assets. This is because a gift of corporate stock conveying a partial interest in the underlying assets may sometimes be valued at a lower value for gift tax purposes than the same fractional interest in the underlying assets. This lower valuation is allowed when the stock reflects a minority interest in the corporation and is called a minority discount.

If farmland is a sufficiently large portion of a farmer's estate, the estate tax can be calculated using a special "use" value for farmland rather than its full market value. This special use value is computed under a formula which, during the 1970s, reduced values for estate tax purposes to somewhat less than 50 percent of market value. Although the maximum reduction in farmland taxable values allowed is limited to $750,000, this special use valuation technique can produce significant savings in estate taxes. Because the estate tax rates are progressive, the tax-savings increase as estate values rise.

In addition, farm and other business estates are entitled to an extended time over which to pay the estate tax. Payments need not start until nearly six years after the death and the tax can be paid in ten equal annual installments. During this time, interest on estate taxes due on the first $1 million of estate value accrues at four percent, a rate well below market interest rates or the interest charged on other tax liabilities.

Labor and Employment Taxes

The Federal tax system imposes two taxes on wages up to certain maximum amounts. The Social Security tax is imposed equally on the employer and the employee; it is also imposed on the business profits of the sole proprietor and on partnerships. Contributions for unemployment insurance are exacted from any employer who, in either the current or previous year, employed ten or more workers at any time in each of 20 or more weeks in the year, or paid $20,000 of wages in any one quarter. If neither of these thresholds were reached in the previous year, there is no liability for these taxes in the current year until one of them is reached. However, once the threshold is reached, all wages for the year, including those paid earlier in the year, are subject to the tax. Thus, wages paid in October can cause a tax to fall on wages paid earlier in the year. As a result, the marginal cost of wages just over the threshold can, at least in one year, be quite high.

Frequently, an employer is also required to contribute to workers' compensation funds. Qualifying criteria and the level of contributions vary from state to state, but they are often significant. These levies not only impose financial burdens, but add to recordkeeping requirements as well. Many farmers and business people may not consider the cost of the tax as onerous as the cost of maintaining such records.

Property and Other Taxation

Property Tax The property tax is one of the most unpopular taxes in rural communities. Another form of wealth tax, the property tax is a primary source of revenue for local governments. The tax falls heavily on real estate, although in some jurisdictions a property tax is imposed on both tangible and intangible property. The dominance of real estate in the asset structure of farm production makes the property tax particularly burdensome for farmers.

The incidence of changes in the property tax in agriculture is borne by the landowner; lower property taxes result in more net income and higher property values, and higher taxes result in less income and lower values. Higher property taxes increase the cost of owning land, and thus encourage more efficient use of land and discourage the holding of land for speculative purposes. However, around metropolitan areas farmland values are influenced by development potential and are consequently much higher than can be justified based on agricultural productivity. Thus, property taxes based on market value may also be very high and may result in high costs of production and pressures to sell the land to developers. To reduce the burden of such high taxes, many state and local jurisdictions have implemented "use value" assessment procedures whereby agricultural land is assessed at its value in current use rather than at its value for future development potential. Such assessment procedures may delay the conversion of agricultural land to urban and development uses and encourage more orderly urban growth and expansion. However, if use valuation is followed, the lower taxes may simply result in higher property values.

Property taxes affect individual investment behavior and decisions about business expansion and location. Permanent improvements to farm property (i.e., buildings, silos, storage facilities) and plant expansion will typically result in higher property taxes, thus reducing incentives to construct such facilities. Some state and local jurisdictions have low property taxes or will provide tax relief to encourage industries to locate or to construct new facilities in a particular location. In many cases, the tax relief given to encourage the location or expansion of a business shifts additional tax burdens to other businesses, property owners, or residents. Unless the additional economic activity of the new or expanded business generates other benefits for these surrounding residents, a permanent redistribution of economic costs and benefits may result from such plans. This issue may be particularly important in rural communities where the number of property owners available to share the "tax subsidy" is limited. The relative size of the property tax burden and the various provisions for tax relief often affect farmers' and agribusiness peoples' willingness to expand and invest in improvements, and property taxes frequently influence the location decisions of agribusiness firms.

In addition to the complaints about the burden of the property tax, a common criticism is directed toward the inequities in the assessment process. Determining the value of property that has not been subjected to market forces is an inexact science at best, and many jurisdictions allow assessments at values substantially lower than market value. Furthermore, assessors are frequently politically chosen or appointed; they often have limited training and little incentive and support to implement the assessment process as completely and accurately as possible.

Sales and Excise Taxes The sales tax is a commonly used source of revenue for state and local governments. The tax is a variation of the consumption tax in that it is applied to the value of products sold. It is collected from consumers by retail firms and remitted to the government. Much of the debate about the sales tax relates to its equity features. Most analysts argue that it is regressive with those having lower incomes paying a larger proportion of their income in tax. Because of concerns about regressivity, some "necessities" such as food are exempt from sales taxes in many state and local jurisdictions. Excise taxes are another form of consumption tax with a focus on specific products such as tobacco, liquor or motor fuels. They are typically collected from the user by the manufacturer or distributor.

Sales and excise taxes are typically "passed through" to the consumer of the product. Thus, product prices are increased and consumption of the taxed product reduced. For agriculture, the major impacts of sales or excise taxes are in the tobacco and wine industries where such taxes increase prices and discourage consumption. Different sales or excise taxes on, or exemptions for various products may also influence prices and purchasing patterns, and consequently affect production organization and investment behavior. However, the actual impact appears to be small. Retail merchants in the agribusiness sector who must collect sales taxes also incur additional accounting and administrative costs in the collection process.

Tax Shelters in Agriculture

Investments taxed under preferential rules, such as the special income and estate tax rules for farmers noted earlier, allow the creation of tax shelters. This tax shelter characteristic affects not only the total financial return from such assets, but may also affect the patterns of asset ownership. Tax shelters lower federal tax revenues and thus require revenue increases elsewhere or contribute to government deficits. The revenue losses due to the special farm tax rules of expensing of outlays and capital gains were estimated for the 1980-82 years as:

Fiscal Year	Expensing of Outlays	Capital Gains
	(Millions of Dollars)	
1980	430	385
1981	475	405
1982	545	425

Source: Congressional Budget Office

Hanson and Eidman have estimated the "tax expenditures" (reduced tax revenues because of special provisions such as investment tax credit, rapid depreciation, etc.) on a per farm basis for the 1967-72 and 1973-78 periods. They found that tax expenditures increased from $2,430 per farm to $8,590 per farm between the two periods; in real terms, tax expenditures increased by more than 100 percent. Total tax expenditures tended to increase as a declining function of sales during the 1973-78 period. Tax subsidies significantly altered the progressivity of the rate structure and effective tax rates were only marginally higher for larger farm firms compared to smaller firms. The Hanson and Eidman study clearly documents the substantial size of the tax subsidies received by farmers and the distribution of these subsidies by size of farm.

What Is a Tax Shelter?

While theoreticians have argued about the precise definition of a tax shelter, the concept is most easily grasped by thinking of tax deductions as "assets" having a valued equal to the reduction in taxes that they will produce. These "tax system assets" are obtained by reporting the deduction on a tax return and thus eliminating some tax liability. From a tax perspective, income is viewed as a "liability," since a tax will be imposed when it is reported on a tax return. The tax accounting rules allow the farmer some freedom in choosing the time to report deductions or to report income. Thus, farm investors would obtain the largest tax-system asset by reporting deductions as early as possible and against income otherwise taxable at high rates. They would have the smallest tax liability by delaying the reporting of income as long as possible and reporting it so as to bear tax at low rates, particularly as long-term capital gains. Proper management of these tax-system assets and liabilities produces a financial benefit that enhances the commercial profits produced by the investment.

Because of the tax shelter potential, high income individuals in particular have strong incentives to report deductions as early as possible, to delay reporting income as long as possible, and to convert ordinary income to capital gains. The returns from actions taken to mismatch income and costs are directly related to the tax bracket of the investor. A high-bracket taxpayer and a low-bracket taxpayer may earn the same commercial return from a tax sheltered farm investment, but the after-tax returns will be greater for the high-bracket taxpayer. Ownership of assets slowly gravitates to those who obtain a greater return and thus can pay the most for them. Thus, over the long-run, ownership of tax shelter assets is concentrated in the hands of the high-bracket taxpayers. The tax shelter means the most to those with the highest taxable income, whether the income is produced on the farm or elsewhere.

Recognizing the distortions attributable to tax sheltering, tax reform efforts during the 1970s were dedicated primarily to closing "loopholes" and ending preferences enjoyed by particular groups. Provisions to eliminate the current deductibility of capital expenditures were used to stop tax-motivated investments in agricultural crops such as citrus and almonds. At the same time, the tax advantages of breeding livestock were reduced by increased holding periods to qualify for capital gains treatment, and recapture of excess depreciation. However, investor interest simply shifted to other agricultural enterprises; grape and walnut acreages increased and cattle-feeding syndicates flourished. The Tax Reform Act of 1976 curtailed the syndication or packaging of agricultural tax shelter investments for sale to nonfarm investors, but individual high income investors continued to reap tax advantages from agricultural investments.

Impacts of Tax Shelters

Studies of the tax shelter potential in agriculture suggest that tax policy has influenced buying patterns and exerted upward pressure on the price of farmland in particular. This pressure occurs because land provides an ideal tax shelter. The return gained by the appreciation in land is not taxed until the property is sold. If the land is held until death, this return is exempted. Carrying costs in the form of interest and property taxes are fully deductible from taxable income and may offset income from other sources. In essence, income taxed at low rates, or perhaps even exempt from tax, is combined with fully deductible cost. This is the classic tax shelter.

Under the Tax Reform Act of 1976, farmland became an estate tax shelter as well. As a consequence, farmland has become more attractive to those who could use an estate tax shelter and afford the land. Preferences in the income and estate tax laws thus have created a demand for farmland that would not exist without them. And, this demand is most likely to come from those in higher tax brackets who can exploit the tax shelter that land provides.

At the same time, the provisions of both the income and estate tax laws contain features that tend to restrict the supply of land offered for sale. For the income tax, the exemption from tax of gains on property that passes at death encourages the holding of land until death. For the estate tax, the ownership requirements that must be met to qualify for the estate tax preferences discourage sales both before and after death.

The greater demand for land and the restriction of its supply have kept upward pressure on the prices of farmland. As the result of tax policy, land is less likely to be sold. Ownership of land is more likely to be passed from generation to generation within the same family irrespective of whether the owners are or are not farmers. Such a policy likely encourages absentee ownership. In order to qualify for estate tax benefits absentee owners are frequently required to participate in some farm management decisions, but they need not reside on the land or be responsible for daily management and operations.

These tax shelter aspects also have encouraged the growth of individual farm firms. Cash accounting allows farming to be a tax sheltered industry. So long as there is other income that would be subject to tax except for the tax shelter, taxpayers in higher tax brackets have more funds for growth and expansion than without the tax shelter. Furthermore, the advantage of cash accounting is augmented if some of the income produced through deductions can be reported as a capital gain, which is taxed at a lower, preferential rate. The greatest advantage goes to the highest bracket taxpayer. An incentive also exists to combine nonsheltered taxable income (for example, a salary) with sheltered income since a tax shelter has little benefit without other income to shelter.

Tax policy and the sheltering potential have also impacted production and prices of farm commodities as well as management practices. The tax law provides incentives for development of orchard crops and the expansion of dairy operations in spite of surpluses and excess production. Studies have explicitly linked favorable tax treatment to increased production and lower prices in the citrus and almond industries. Culling practices in the hog, beef and dairy industries are altered in part to take advantage of the favorable taxation of capital gains.

Tax provisions also affect the choice of the legal form of a business. Frequently, lower overall tax rates are achieved by incorporating the farm. Such lower tax rates provide more after-tax funds for expanding the operation. As noted earlier, if the corporate earnings are paid out to shareholders, they will draw a dividend tax. Yet, if they are kept at the corporate level without being used in the business, they may draw the penalty tax on accumulated earnings. Both horns of this dilemma are avoided by using the corporation's earnings to expand the operation. The estate tax preferences noted earlier also encourage continuing a business over many generations and expanding it.

Policy Issues

A number of tax policy issues will be debated in the

next decade. Part of this debate will focus on the general concept of taxation and the implications of alternative tax systems on various economic sectors including agriculture. Some of the debate will focus on agriculture, especially the impact of tax provisions on incentives for capital investment and tax sheltering. The following discussion will identify some of these issues and summarize relevant research results.

What to Tax?

A fundamental issue in tax policy is what concept of taxation should be adopted--taxation of wealth and/or property, or taxation of annual flows of benefits from that property and other economic activity. In the U.S., federal and state taxes are leveled primarily on annual flows in the form of income or sales taxes (with the exception of the estate and gift taxes which are a minor source of revenue), and local taxes have focused on wealth or property. This relative focus and the relative tax burden on wealth compared to annual flows is not expected to change in the future unless the proportion of public services provided by local, compared to state and federal government agencies, changes significantly. The dominance of the property tax as a source of revenue for local governments places a relatively high burden on agriculture compared to other economic sectors. Thus, increased demand on local governments for services without changes in the local tax base or taxation system would increase the relative tax burden of agriculture.

A second tax issued is what annual flow should be taxed--income, consumption, sales or "value added". Most of this debate focuses on the income, consumption and valued added approaches. In the U.S., the primary focus of the federal government is on income tax, but recent concerns about savings behavior and incentives provided by a consumption tax (a tax based on expenditures rather than income) to encourage savings has stimulated interest in possible changes in the taxation system. One basic concern with the consumption or expenditure tax is its regressivity unless certain expenditures such as food are exempt.

A value added tax is really a sales tax imposed on the increment of value added at each stage in the production or manufacturing process. It is also regressive and may be complex to administer. Some have argued that the "hidden" nature of the value added tax (the tax burden becomes buried in the price of the final product) does not provide the discipline associated with the direct payment of taxes. Thus, the public is less aware of the true nature of the tax burden. The implications of these three different concepts of taxation for agriculture have not been evaluated extensively. However, if food products and agricultural production were exempt from a consumption or value added tax as is frequently suggested, the tax burden on agriculture may be reduced compared to the taxation of income.

Tax Base and Rates

If income is the base in assessing taxes, a further set of policy issues surface. First, how is income to be measured and second, what rates should be applied to the income base? Three dimensions of the tax rate structure are of interest: the overall rate level, the concept of progressivity, and indexing of rates to inflation.

Progressive income taxation has been part of the U.S. tax system since its inception, although justifications for it have varied. Proponents of a progressive system argue that by having the rich contribute proportionally more, government revenues are raised at least "cost." "Rich" persons suffer less with the loss of a dollar to taxation than "poor" persons, given diminishing marginal utility of income (Vedder and Frenze, 1983; Blum and Kalven; Minarik, 1982a). Progressive income taxes may compensate for regressivity of other taxes in the overall system. Higher tax rates at high incomes can be justified if more benefits from tax-funded programs accrue to higher income taxpayers.

Progressive income taxation is opposed by those who believe that workers are entitled to the "fruits of their labor" and that progression cripples economic efficiency. Simon claims that proponents base their support of the system largely on ethics and political realities, not economic considerations. The more progressive is an income tax system, the greater are the disincentives to work, save, and invest, causing misdirection of resource use (Blum and Kalven; Simon; and Blum).

The progressive tax system incorporates automatic

stabilizers; that is, progression results in changes in average effective tax rates without official intervention. Taxes rise more than proportionally when nominal incomes increase and fall more than proportionately when incomes fall. But, inflation can create undesired inequities in the tax system if nominal incomes are used for tax calculations over a period of years.

Recent proposals have suggested the replacement of progressive tax rates with a "flat rate tax." A pure flat rate tax would have two possible marginal rates: zero, if credits exceed the tax liability on income, or otherwise, the flat rate. Some elements of progressivity are retained in the tax system if exemptions are allowed with the flat tax; high income persons have large average shares of their income taxed instead of having higher marginal tax rates.

Tax liabilities with a flat rate system would be less sensitive to inflationary effects and bracket creep would be eliminated. Income reallocation between years and income averaging would become unnecessary since the tax rate would be the same from year to year (Blum and Kalven; Minarik, 1982a, b). If the corporate rate schedule were also replaced by a flat tax, incentives to incorporate might be reduced. In a study by the Commission to Revise the Tax Structure, a flat rate tax schedule in conjunction with a broader tax base was one of the variants evaluated. The flat rate schedule led to a decline in the progressivity of the tax system with effective tax rates increased for incomes below $25,000 and decreased for incomes above $25,000.

In Slemrod and Yitzhaki's recent article "On Choosing a Flat-Rate Income Tax System," an optimization algorithm was used to find a flat rate tax system which approximates the current distribution of tax burdens. They evaluated systems which minimized either the sum of the changes in the tax liabilities or else the sum of changes in average tax rates. They conclude that middle income families (those with $20,000-$50,000 in income) would bear the brunt of the shift in tax burdens associated with a flat rate.

Minarik (1982b) found that a flat rate tax of 11.8 percent on adjusted gross income with long-term capital gains included, no itemized deductions, no zero bracket amount, and no personal exemptions had significant income redistribution potential. A system which involved less dramatic changes from the present was one with an 18.7 percent tax on adjusted gross income with long-term capital gains included in full, no itemized deductions, a $1,500 personal exemption, and a $3,000 ($6,000) zero bracket amount for single (joint) returns. He concluded that a flat rate income tax would lower the average and marginal tax rate at high incomes and almost certainly increase the average tax rate for middle income individuals. He points out that positive incentives for work, savings, and investment due to lower tax rates for those with high incomes could be offset by disincentives for those with middle level incomes.

Minarik also suggested that "good old fashioned tax reform" might be a viable alternative. By incorporating numerous base broadeners, the tax rate schedule could be lowered and flattened. A 14 percent first bracket rate with low income relief could give an estimated 75 percent of the taxpayers a flat rate tax without redistributing income. Two or three additional brackets could be used to approximate the current tax burden on high incomes while simplifying the system. Some of the existing tax problems--bracket creep and other inflationary side effects, and savings and consumption disincentives--might be reduced though not eliminated.

Tax reform that included base broadening measures would result in changes in tax provisions enacted in the past to aid farmers. Broadening the tax base means including more of personal income in tax calculations. Actual farm expenses would still be deductible but existing tax concessions would be reduced and incentives removed, promoting simplicity, efficiency, and horizontal equity. Political feasibility would likely temper selection of provisions to broaden the tax base.

Tax treatments that especially benefit agriculture emanate from three major sources: (1) a choice of accounting methods, cash or accrual, (2) options as to the method used to write-off capital expenditures, and (3) the right to receive capital gains treatment on qualifying livestock. Under current laws, farmers are able to sell livestock raised and used as breeding stock and treat gains from the sales as capital gains, which are taxed at a lower rate. Costs of developing some farm assets--certain conservation expenses, costs associated with caring for orchards and vineyards, and costs of raising livestock to maturity--may be deducted in the year they are incurred or paid. Tax liabilities can also be reduced through "do-it-yourself" averaging, given the flexibility in reporting income and expenses of cash accounting. Under a flat rate system there would be little incentive for adjusting income and expenses at year end, since the tax rate would be the same from year to year, regardless of the income level.

The implications of a flat tax and base broadening on farm firms and agriculture has received some research attention. Analysis of the micro impacts of a 20 percent flat tax for 1984 show that effective tax rates tend to rise slightly for lower income farmers and fall substantially for the higher income producers (Doye and Boehlje). Because of their higher incomes, representative hog producers tend to benefit relative to grain producers under the flat tax on the current base. For the low income, low equity grain producers, changing to a flat tax would not matter because they have no tax liability under either set of tax rules.

Simulation of the flat tax with a broadened base (i.e., fewer exemptions and deductions) shows the same pattern as the flat tax simulations using the tax base under 1984 tax law. The larger, higher equity and higher income farms show effective tax rate reductions, while on the smaller, lower equity and lower income farms the tax rates are greater than or equal to those under ERTA provisions. Compared to the results using the 1984 tax base, hog producers have higher effective tax rates with a broadened base and grain producers show lower effective tax rates. This higher rate for hog producers results from the elimination of credits and depreciation reductions that are relatively more important to hog farmers. Grain farmers do not benefit significantly from the capital gain exemption, and benefit relatively less from the investment credit and ACRS provisions compared to hog farmers. Thus, the elimination of exemptions and credits as part of the base broadening is relatively unimportant. Moreover, the lower statutory tax rate with a flat tax also results in lower effective rates.

Hardesty and Carman evaluated the 1984 Treasury Department tax simplification proposals. These proposals include lowering and consolidating the tax brackets to 15 percent on the first $31,800 of taxable income, 25 percent on incomes between $31,800 and $63,800, and 35 percent for amounts exceeding $63,800; indexing the tax rate, interest deductions and interest earnings to inflation; elimination of capital gains tax treatment; lengthening depreciation schedules; and eliminating investment credit. Their results indicate that machinery purchases would decline because of the elimination of investment tax credit and extending depreciation periods. When all the provisions are combined, land purchases are encouraged relative to machinery investments. According to their analysis, tax simplifications, as reflected in the 1984 Treasury proposal, would result in higher net worth, increased savings, additional land expansion and debt utilization to finance land purchases, and decreased total taxes.

Some key issues in the simplification and flat tax debate relate to the rate level to use and the relative effects on various sectors of the economy because of differences between sectors in effective rates and the opportunity to use various deductions and exemptions. Sisson considered whether farmers have a significant tax advantage over the general population. From his examination of farm-nonfarm tax burdens, he concludes that farmers, and large farmers in particular, have substantially lower tax burdens than nonfarm taxpayers under progressive rates of the 1960s and 1970s. A 20 percent flat tax rate is higher than the average federal tax burden estimates of Sisson, regardless of the definition of farmer used and whether overall tax incidences are assumed to be regressive or progressive. Sisson's results imply that many farm families would experience increased taxes under a flat tax proposal if the cash accounting system and other deductions and exemptions were eliminated.

Sisson also concluded that farm income distributions are more skewed than for the general population with farm incomes concentrated at low levels. Data from Agricultural Statistics and the Statistical Abstract of the U.S. substantiate his conclusion; farm families are more concentrated at the lower income levels than are families in the general population. Thus, adoption of a flat rate tax could contribute to increased skewness in the distribution of farm incomes, since high income units would receive a tax break while low income units would not. Furthermore, if the flat tax were accompanied by a broadening of the tax base through the elimination of exemptions, credits and deductions, intensive livestock producers and the growers of perennial crops would lose relative to other farmers because they enjoy significant tax preferences under 1984 law.

Investment Incentives and Tax Sheltering

As noted earlier in this chapter, tax provisions have a significant effect on investment behavior. Some have argued that the incentives for capital expenditures in agriculture provided by the tax law may be beneficial to farmers in the short run, but result in excess capacity, lower incomes and increased competition from nonfarm investors in the longer run. This debate about the impact of tax law on agriculture focuses primarily on the cash accounting system, the treatment of capital gains, and the write-offs and tax credits on capital expenditures.

Cash Accounting Probably no issue in the reporting of farm income has received as much attention as the use of cash accounting by farm operators. Volding and Boehlje simulated the impact of different accounting procedures on six different types of farms in two different size categories as measured by farm receipts. The objective of the analysis was to choose an accounting system that maximized the discounted after-tax income over a five year period. Three accounting systems were compared for the various farm situations: the accrual system, the cash system with maximum cash adjustments, and the cash system with optimal cash adjustments. Cash adjustments included expense items that are prepaid, as well as income items that are postponed.

The results indicate that for all farm sizes and types, the cash method of accounting with optimum adjustments is preferable to the other two accounting systems. Over the five year period this method of accounting produces more total dollars for after-tax income, consumption, and change in net worth. It also results in a more rapid growth rate. Furthermore, larger farms in each enterprise type receive a higher payoff from the cash method with optimum adjustments compared to their smaller counterparts when after-tax income is considered. One major reason for this is that larger farms have more earned income and consequently have higher marginal tax rates; one dollar in additional cash adjustments saves more income from taxes when the marginal tax rate is higher.

Large grain farms benefit more from the cash method with optimum adjustments than do other large enterprise types. Grain farms have the potential for much higher cash adjustments than other types of farms because many of the inputs (such as seed, fertilizer, and chemicals) are purchased off the farm and can be prepaid, and all production of grain can be held from sale. In contrast, a beef cow-calf farm, for example, can also hold its production from sale. However, it has a much lower net income than does the grain farm, even though both kinds of farms are within the same gross sales size classification.

Other studies have shown similar results. Bryant, LaDue and Smith demonstrated that use of the cash accounting system on dairy farms resultd in a substantial increase in firm growth over time. The exploitation of the cash accounting rules in the cattle feeding industry during the late sixties and early seventies has been well documented by Meisner and Rhodes. Their work suggests that the tax advantages of the cash accounting system combined with the limited partnership investment vehicle were major factors in the development of the Southern Plains cattle feeding industry during this period.

Cash Accounting and Capital Gains Combining cash accounting with the special tax treatment of capital gains is especially attractive for breeding herds. The application of cash accounting and capital gains provisions to swine farrow-finish and feeder pig enterprises was analyzed by Duffy and Bitney. In each enterprise, results were calculated for a strategy using all gilts for pig production, and then for a strategy in which sows were kept for four litters. Under the all-gilt strategy a much larger proportion of total sales will qualify as long-term capital gain.

Costs and returns from one farrowing of 32 sows under the four-litter strategy resulted in more before-tax profit than the all-gilt strategy. When income taxes are considered, the after-tax profits from the two breeding herd replacement strategies are exactly equal at an average tax rate of 39 percent. The after-tax profit advantages at tax rates above 39 percent are not large, particularly when the nonquantifiable factors and the 50 percent maximum tax rate on earned income are considered. A similar comparison for full-time feeder pig production yields quite different results. The analysis shows that after-tax profits for the all-gilt breeding herd replacement strategy are greater over the entire range of feeder pig price levels. These differences in after-tax profits between the all-gilt and the four-litter strategies are significant, both absolutely and in proportion to the total. The all-gilt strategy produced nearly twice as much after-tax profit as the four-litter system at a $30-per-head feeder pig price. But, the all-gilt program requires large numbers of replacement breeding stock which may present a problem for the producer who sells pigs at 40 pounds. If replacement gilts are bought, the capital gains advantage associated with raised breeding stock is lost.

A study of the differential tax treatment of ordinary income compared with capital gains on the optimal enterprise organization and management practices for crop-hog farms in Georgia suggests similar results to that noted above (Reid, Musser, and Martin). The optimal farm organization was compared on a before-tax and after-tax basis. Inclusion of income taxes in the analysis resulted in the hog enterprise being a more dominant part of the farm operation, particularly for larger farms, along with heavier culling of sows and a larger proportion of gilts in the breeding herd. Another analysis found an incentive for crop farms to move toward production of animals such as hogs when capital gains tax provisions are incorporated in the analysis (Musser, Martin, and Sauders). Overall returns are increased by deducting animal development costs against the crop income, and then reporting a significant proportion of income from the animals as long-term capital gains.

The effect of various Federal tax provisions on dairy farms has also been analyzed (Bryant, LaDue and Smith). By use of cash accounting and treating livestock profits as capital gains, the tax bill on the representative dairy farm over the 20 year period was cut in half compared to accrual accounting and reporting livestock sales as ordinary income. The increase in net worth by using these provisions was 55 percent greater. The differences could be even larger if the taxes on liquidation at the end of 20 years could be avoided. The study assumed a disposition of property through sale. If, however, the property was retained until death, it would acquire a basis equal to the value at death, and no recapture of investment credit and depreciation would occur.

In general, these and other studies suggest that the cash accounting system combined with capital gains provisions encourages farm expansion, changes in management practices, and non-farm investment in agriculture to shelter taxes. The basic policy issues are: (1) should the cash accounting system be limited only to farmers, and (2) should more restrictive limits be placed on the size of farm that can use the cash accounting procedures?

Capital Gains, Interest Deductions and Land Prices

The special tax treatment of capital gains, the interest deduction, and the computation of the basis in real property appear to affect the rate of wealth accumulation, the ownership patterns, and prices of farm real estate. The effect of capital gains on wealth accumulation and land values was evaluated by Boehlje using a dynamic business simulation model. Two different sets of rates of return and price appreciation were specified for analysis. The first set included a four percent cash rate of return on real estate and an eight percent rate of price appreciation; the second set included an eight percent cash return and a four percent appreciation rate. The total return was 12 percent before taxes in both cases. The only difference was in the mix between current cash income taxed as ordinary income and appreciation taxed as capital gains. With the lower cash returns, current consumption also was reduced because consumption was specified as a function of cash income.

The implications of these different rates of return and appreciation for different sized Iowa farms illustrate the importance of the tax treatment of capital gains and ordinary income. For a $1 million Iowa farm, the after-tax ending equity was 16.5 percent higher with the high appreciation rate/low cash return rate assumption. For the $3 million farm, the after-tax ending equity was 27.4 percent higher with the higher appreciation rate. Thus, the preferred tax treatment of capital gains results in more wealth accumulation when a large portion of the total return is appreciation rather than current cash income. Furthermore, in this circumstance, the benefits of the differential tax treatment for capital gains were larger for the larger farm. Because the larger farm has a higher tax bracket, the tax savings from deferring the payment of taxes is larger when more wealth accumulation accrues as capital gains.

This study also evaluated the effect on land bid prices of the differential tax treatment for capital gains compared with ordinary income, along with various other parameters such as leverage and holding period. Bid prices declined as the rate of current return increased and the appreciation rate decreased; the major reason for the decline is the tax treatment afforded the two sources of return. A higher price can be paid for real estate that yields a higher proportion of its return in the form of capital gains. Moreover, the tax benefits of capital gains are larger for people in higher marginal tax brackets.

The marginal tax rate has an interesting impact on the bid price. With 20 percent leverage, the bid price declines as the tax rate increases, assuming the holding period and return-appreciation parameters are held constant. With a combination of high leverage and low current return/high appreciation rate, the bid price goes up as the tax rate increases. However, if the current return is high and the appreciation rate low (again assuming high leverage), the bid price goes down or increases only slightly as the tax rate increases. Thus, if the earnings are taxed on a current basis as is the case with the higher current return/low appreciation rate assumption, a higher tax rate results in a lower after-tax income and a lower bid price for land. If, on the other hand, the earnings accrue primarily as capital gains, and the purchase is highly leveraged, the tax benefits produced by the interest deduction plus the deferred taxation of capital gains allow a higher bid price as the tax rate increases.

Evaluation of leverage found that increased leverage has a relatively small impact at low tax rates. Assuming a constant cost of equity capital as the tax rate increases, the impact of leverage becomes more significant. In higher tax brackets, the differences between the bid prices for high compared to low leverage are larger with high appreciation/low current returns compared to low appreciation/high current returns. These differences occur because of the larger value of the interest deductions with higher tax brackets.

This study and other analyses of capital gains provisions raise a number of policy issues. Should capital gains be taxed differently than other sources of wealth accumulation (i.e., income)? Does the availability of capital gains tax treatment for some assets distort the investment pattern in agriculture? What would be the impact of eliminating capital gains tax treatment for qualified sales of breeding livestock? Is capital gains tax treatment needed to encourage investment in agriculture?

Analysis of ERTA and TEFRA The Economic Recovery Tax Act of 1981 (ERTA) and companion legislation in the Tax Equity and Fiscal Responsibility Act of 1982 (TEFRA) were major changes in the U.S. tax system that affect almost every farmer. The legislation reduced income and estate tax rates, streamlined depreciation methods through the Accelerated Cost Recovery System (ACRS), and changed investment tax credit and estate tax rules. Analyses of these changes in the tax law are useful in asessing the effects of additional proposals to reduce the tax rates and alter the credits and write-offs associated with capital investments.

Simulations of representative hog, grain, beef and grape farms suggest tax liability reductions from the ERTA changes in tax law for all producers evaluated throughout a 10-year planning horizon (Lowenberg DeBoer and Boehlje). ERTA provisions result in increased firm expansion and net worth growth on all farms analyzed. In simulations using representative initial financial condition and depreciation schedules, net worth growth was from 4 to 17 percentage points higher under ERTA provisions than in those using the pre-1981 tax rules. Larger, higher income farms tend to benefit from ERTA more than the smaller, lower income operations. Farms with higher off-farm income or larger amounts of equity in the financial structure reap larger benefits from ERTA as well. The ACRS rule which allows specialized livestock facilities to be depreciated over 5 years, instead of 15 or more, is of special importance to hog producers. Concord vineyards also benefit from the ACRS system, but because income growth is relatively low for vineyards, they are less above to use the ERTA provisions than the hog farms.

Simulations which used ERTA tax rates, but left all other tax provisions unchanged, showed that the tax rate cuts dominate all the other ERTA reforms. The effects of changes in depreciation rules are relatively minor. For beef and grain producers who previously used accelerated depreciation methods, ACRS shows little benefit. The changes in investment credit rules in ERTA and TEFRA appear to have little effect on the actual investment credit on farms. The elimination of investment credit has about the same effect on simulations under the old and new tax rules, if the expense method of depreciation is not used. Simulations which use the option to expense some investments do show a large drop in investment credit use because expensed items do not earn the credit. Simulations of the illustrative farms organized as corporations show relatively small benefits from ERTA and TEFRA compared to the benefits for sole proprietorships.

Estate Tax Provisions

Estate taxes have been perceived by farmers to represent an inordinate tax burden which could destroy the family farm because the heirs must frequently sell part of the farm to pay the taxes. Based on this perception, significant changes have been made in recent years to reduce the estate tax burden for farmers. Most discussions of this issue have emphasized the impacts of changes in the law on individual farms; the structural implications for the sector are not well understood or documented. With declining resource values in agriculture, one policy issue is whether scheduled reductions in effective estate taxes should be maintained. If so, what are the implications of some of the special provisions available to farmers?

Boehlje evaluated the outflow of funds to pay estate taxes and to defray estate settlement and liquidation costs, and determined the incentives provided by the Federal estate tax laws for farmers with different characteristics to change their land ownership status (buy, sell, transfer, or lease). To accomplish these objectives, illustrative farms were selected with different size, asset composition, tenure, financial structure, and other characteristics and were evaluated with a dynamic estate and business planning simulator. The analysis was completed prior to the recent phased in reduction of effective estate tax rates through increased credits, but the general implications are still valid.

The results indicate that the absolute value of the tax savings from using special estate tax provisions available to farmers such as special-use valuation and installment payment of taxes are generally largest for farms with the largest net worth. These are also the farms that have the largest relative and absolute tax burden without the use of these provisions. In contrast, these special tax provisions do not have as much absolute benefit for the smaller illustrative farms.

Tax savings from special-use valuation are not proportional to estate size. The percentage tax reduction from this provision is substantially lower if the estate is large enough to exceed the maximum allowable reduction in estate valuation from special-use valuation. Thus, the tax savings from special-use valuation increase in absolute magnitude but decrease as a proportion of taxes due as estate size increases. For smaller estates that can use the marital deduction and unified credit to fully offset estate taxes, special-use valuation yields little estate tax savings and may increase income taxes at a subsequent sale, because the use value establishes the tax basis for the property. Thus, at a later sale, the land would have a larger capital gain and capital gain tax if use valuation, rather than fair market valuation, is used to value the property at death.

The tax savings from special-use valuation also depend on the relative proportion of land in the estate and the quality of the land. A larger proportion of land increases the amount of the estate that can receive special tax treatment. Moreover, higher valued land appears to receive a larger discount from using special-use valuation, and thus results in more tax savings compared with lower valued land. The relative and absolute tax savings from special-use valuation are also substantially larger when the farm includes more assets and more debt but the same net worth. This larger savings occurs because the leveraged farm includes more land assets which qualify for special-use valuation.

The option to pay taxes in installments allows the heirs to use the earnings from the farm and other sources of income during the 15-year period following death to pay the taxes. The tax savings from installment payment of taxes remain approximately proportional with increases in farm size until the taxable estate reaches the size where the interest rate increases from four percent to the regular rate on unpaid tax; beyond this size the tax savings decline. Since the installment payment option reduces the need for liquid funds to pay taxes, the installment payment of tax provisions may have a greater effect on the continuity of the firm and help to maintain the size of the farm after the parents' death than special-use valuation.

Conservation Practices and Tax Law

The Internal Revenue Code permits certain landowners to deduct their share of the cost of soil or water conservation projects as an operating expense. These deductions are limited to 25 percent of the gross income from farming in any given year, but the balance of expenses may be carried over to future years. In 1969, a recapture provision was added which required that if land that produced a soil or water conservation deduction is sold less than ten years from acquisition, a part or all of the gain attributable to the deduction is recaptured as ordinary income.

A 1979 study found that these tax provisions and other government policies taken together have a significant impact on the adoption of soil and water conservation practices (Boggess, et al.). Specifically, the effectiveness of interest or capital subsides for terraces largely depends on a farm's income and marginal tax rate. With the low income generated by farms on the more erosive soils, terracing was not used in the absence of legal restrictions on soil losses. Thus, terracing subsidy programs had little effect on acres terraced. However, when livestock was added, acres terraced increased dramatically, even without a cost or interest subsidy, because of the higher income and tax incentives that permit deducting such expenses up to a limit of 25 percent of taxable farm income. For farms on less erosive soils, where incomes were higher, terracing was included as part of the base plan, with the amount of terracing depending on the level of income, and thus the potential tax savings from deducting the terracing expenses.

With growing concern about the lost productivity and environment degradation from soil erosion, public policy approaches to encourage adoption of conservation practices are receiving increased attention. Tax law may be an effective vehicle to encourage conservation. Similar provisions that allow deductions for land clearing and destruction of woodlands may not have a similar social benefit.

Aggregate Effects

Much of the debate about changes in the tax laws has focused on short run, firm level implications; the longer run industry response has not received adequate emphasis. A fundamental issue is even if farmers receive short term benefits from changes in tax laws, what is the longer run impact given changes in investment patterns, supplies, and prices?

The industry effects of changes in tax provisions are difficult to ascertain. Agricultural producers respond to many factors in their investment and production decisions, and their responses often involve significant time lags. The short- and long-run implications of tax law changes may differ as may the individual and aggregate impacts. Analysis of expected industry and macro adjustments that will result from the Economic Recovery Tax Act of 1981 provides a basis for suggesting the nature of some of these effects.

The Accelerated Cost Recovery System (ACRS) and revised investment tax credits can be expected to have differential effect on assets because of different relative changes in tax lives. There also will be a differential impact on industries because of different mixes of capital stock. Gravelle analysed the effects of ACRS on effective tax rates by asset type and industry under two annual inflation rates. This analysis, which considered only equipment and structures, estimates that effective tax rates on a marginal increment of investment in agriculture decreased from 29.5% to 16.7% assuming 6% inflation, or from 34.5% to 22.5% assuming 9% inflation. The ranking of eleven broad industries from highest to lowest in terms of effective tax rates places agriculture fifth prior to and fourth after passage of ERTA. Eight of the other ten industries fared better than agriculture when considering relative decreases in effective tax rates. Despite some omissions and the failure to consider some special tax rules, the results indicate that ERTA does tend to favor capital investment in industries other than agriculture.

Durst and Jeremias concluded that the effective income tax rates for farm capital have declined significantly since 1980 because of a sharp decline in the rate of inflation as well as the marginal tax rate reductions contained in ERTA. Without the significant reduction in the rate of inflation, average tax rates for low income farm investors would have actually increased in spite of the ERTA changes according to their analysis. They conclude that in spite of indexation of tax rate schedules and exemptions, inflation will be the dominant determinant of effective tax rates on farm capital through its impact on the real values of nominal depreciation deductions and tax credits. With lower rates of inflation, effective tax rates will remain below the 1980 levels; double digit inflation would be needed to raise effective rates above those of 1980.

The simulation results noted earlier suggest that ERTA and TEFRA have major distributional impacts within the farm sector. Higher income farmers tend to benefit more than lower income producers. This occurs regardless of whether the higher income comes from large farm business size, greater efficiency, higher equity, or more off-farm income. This may accentuate the trend toward a bimodal size distribution of U.S. farms. Producers with large scale farms will benefit because of lower taxes on farm income. Small scale part-time farmers will benefit because of lower taxes on their off-farm income. The farmer in the middle whose operation is too large to permit much off-farm income, but too small to have a high farm income, will tend to have smaller tax benefits.

For some assets the Accelerated Cost Recovery System substantially increases the present value of income tax deductions when compared to prior methods of depreciation. The increase is primarily a function of the decrease in tax life of various assets due to ACRS; examples include the dramatic changes for trees and vines, tile, and single-purpose structures. Thus, the ACRS provisions in ERTA can be expected initially to have a differential impact on after-tax returns from farm enterprises because of variations in the mix of depreciable assets. Because farm types and enterprise mixes vary by state and area within the United States, ERTA can have significant regional impacts on effective tax rates and after-tax rates of return and capital investment (Boehlje and Carman).

Finally, the dramatic decrease in tax lives for specialized livestock and horticultural operations, trees and vines included in ERTA may also have a significant impact through time. In the vineyard area, investor interest will shift in the short-run from developing orchards to purchasing bearing orchards. Budgeted examples indicate that capital recovery over five years from bearing orchards provides more after-tax income than orchard development, even when development costs are deducted from other income as a current expense (Hardesty and Carman). The price of bearing orchards is expected to increase, with the maximum increase dependent on the marginal income tax bracket of the investor.

Analysis of these longer run supply and price responses to changes in tax policy affecting speciality crops indicates that the short-term tax benefits may not persist over time. Legislation in 1969 and 1970 required capitalization of all citrus and almond orchard and grove development costs for the first four tax years after planting. These provisions were sponsored by industry participants who were concerned about the long-term impact on acreage, production, and prices of syndicated tax shelters in these crops. Capitalization requirements significantly increased the after-tax costs of developing citrus groves and almond orchards and effectively terminated the tax shelter advantages of grove development.

Carman's work indicated that the 1969 and 1970 legislation had an immediate impact on new plantings and total acreage of California citrus and almonds. Decreased plantings are reflected in changing bearing acreage, production, and prices over time. The estimated immediate impact of tax reform was to reduce total annual acreage of citrus and almonds. Alternative tax shelter orchard crops--walnuts, avocados, and grapes--increased with tax reform. Requiring capitalization of citrus grove and almond orchard development costs was associated with an immediate decrease in grove and orchard values. The per acre decrease in values was almost three times as large for navel oranges and lemons as it was for almonds. Over-time increased product prices due to tax reform helped offset the initial decrease in values for almond orchards and citrus groves.

Although sector studies like Carman's and others are a significant extension beyond the micro studies summarized earlier, they still may not be sufficiently inclusive to document the total net effects of changes in tax law. Hughes and Adair evaluated direct and indirect impacts of the Economic Recovery Tax Act of 1981 and subsequent amendments. The direct impacts include the reductions in tax rates and the more rapid write-offs for capital purchases which decrease the effective costs of capital and encourage new investment and expansion. The indirect effects occur through the increased government deficit because of lower tax revenues, and the resulting higher interest rates because of increased demands by the public and private sector on the financial markets. The increased demand for durable assets also results in higher prices for those inputs. Hughes and Adair conclude that "given the interest rate sensitivity of net farm income and the effects of higher interest rates and higher prices of durable inputs on the cost of capital, lower tax rates introduced in ERTA may reduce the optimal agricultural capital stock below what might have existed with the old tax laws . . . . Thus, the farming sector may be one of many sectors that show a perverse reaction to the policy of using general tax reductions to encourage investment."

Conclusions

While estimates about the strength of tax policy as a determinant of the course of agriculture are not available, studies completed by the USDA (Davenport, Boehlje and Martin) and others conclude that:

- o Tax policy has exerted upward pressure on the price of farmland.
- o Tax laws have encouraged expansion of individual farm firms.
- o Tax laws appear to impose taxes on labor while allowing tax breaks for capital investments, thus encouraging the substitution of capital for labor.
- o Tax shelter aspects of farm tax laws have stimulated the production of tax-sheltered crops.
- o Tax laws encourage the incorporation of farm and agribusiness firms.
- o The single tax treatment of farm cooperatives is justified for those cooperatives whose activities are a legitimate extension of the business focus of the member-patrons.
- o Farmers frequently alter management practices to take advantage of the tax preferences extended to some farming operations.

- o Except for reducing the consumption of tobacco and wine products, sales and excise taxes have little impact on the structure, efficiency, or organization of the farm and agribusiness sector.
- o Property taxes are capitalized into land values. Tax relief may discourage the sale of farmland for development purposes, but if such relief is indiscriminately applied to farmland with no development potential, it will also increase land values.
- o Differential rates in property taxes and various forms of tax relief will influence farmers' and agri-business peoples' decisions about when to expand and where to locate.

Policymakers will face a number of tax issues in the future. Use of the tax law to provide incentives for changes in economic activity will always be a temptation (for example, to encourage energy conservation, investment in new plant and equipment, and rehabilitation of old buildings). In many cases, long-run and structural effects of such tax incentives may not be as desirable as the intended short-run effects.

To analyze and debate future tax policy, there should be a better understanding of the long-run efficiency, as well as the structural effects, of changes in tax laws. For example, have the investment tax credit and the faster deductions under the ACRS depreciation rules encouraged the adoption of more efficient production methods, or have they simply encouraged over-expansion and capital intensive investments using debt funds? Is the availability of cash accounting in the best long run interest of farmers? Should nonfarmers be able to use farm losses to shelter income from other sources? What are the consequences of property tax relief and reduced farmland assessments for land use and real estate prices? Should cooperatives maintain their unique tax status? Are further reductions in tax rates desirable? Would agriculture fare better in the long-run under a flat tax rather than the current progressive tax rate structure? What are the costs and benefits of indexing tax rates to the rate of inflation? What implications does a value added tax or a tax on consumption have for farmers and agribusiness people? These and other issues will comprise the tax policy agenda in the future.

REFERENCES

Blum, W. J. and H. Kalven, Jr., The Uneasy Case for Progressive Taxation, University of Chicago Press, Chicago and London, 1953.

Blum, W. J., "Revisiting the Uneasy Case for Progressive Taxation," Taxes 60(Jan. 1982):16-21.

Boehlje, Michael, "An Analysis of the Implications of Selected Income and Estate Tax Provisions on the Structure of Agriculture," CARD Report 105, The Center for Agricultural and Rural Development, Iowa State University, Ames, Iowa, October 1981.

Boehlje, Michael, "Taxes and Future Food and Fiber System Structure Performance," The Farm and Food System in Transition, Emerging Policy Issues, Cooperative Extension Service, Michigan State University, 1984.

Boehlje, Michael and Hoy Carman, "Tax Policy: Implications for Producers and the Agricultural Sector," American Journal of Agricultural Economics, Vol. 64, No. 5, December 1982, pp. 1030-1038.

Boehlje, Michael and Kenneth Krause, "Economic and Federal Tax Factors Affecting the Choice of a Legal Farm Business Organization," National Economics Division, Economics and Statistics Service, United States Department of Agriculture, Agricultural Economic Report No. 468, June 1981.

Boggess, William, James McGrann, Michael Boehlje, and Earl Heady, "Farm-Level Impacts of Alternative Soil Loss Control Policies," Journal of Soil and Water Conservation, Vol. 34, No. 4, July-August, 1979, pp. 117-183.

Breimyer, Harold, "The Reagan Tax Plan's Impact on Family Farming," CRA Newsletter, Center for Rural Affairs, Walthill, Nebraska, July 1985.

Bryant, William R., Eddy L. LaDue, and Robert S. Smith, "Tax Reform and Its Effect on the Dairy Farmer," Department of Agricultural Economics, Cornell University, Agriculture Experiment Station, May 1973.

Carman, Hoy, "Income Tax Reform and California Orchard Development," American Journal of Agricultural Economics, Vol. 63, 1981, pp. 165-180.

Congressional Budget office, "Tax Expenditures: Current Issues and Five Year Budget Projections for Fiscal Year 1982-86," Washington, D.C., 1981.

Davenport, Charles, Michael D. Boehlje, and David B.H. Martin, "The Effects of Tax Policy on American Agriculture," Economic Research Service, United States Department of Agriculture, Agricultural Economic Report No. 480, February 1982.

Davis, J. Ronnie and Charles W. Meyer, Principles of Public Finance, Prentice Hall, Englewood Cliffs, N.J., 1983.

Doye, Damona and Michael D. Boehlje, "Flat Tax Rate: Impacts on Representative Farms," Department of Economics, Iowa State University, Ames, Iowa, 1985. Also, forthcoming in Western Journal of Agriculture Economics, December 1985.

Duffy, Michael and Larry Bitney, "The All Gilt Breeding Herd--More After-Tax Profits?" University of Nebraska, Department of Agricultural Economics, Report 77, May 1977.

Durst, Ron L. and Ronald A. Jeremias, "The Impact of Recent Tax Legislation and Inflation on the Taxation of Farm Capital," Agricultural Finance Review, Vol. 44, 1984, pp. 36-42.

Feldstein, Martin, "Inflation, Portfolio Choice and the Price of Land and Corporate Stock," Working Paper No. 526, National Bureau of Economic Research, Inc., Cambridge, Massachusetts, August 1980.

Gravelle, Jane E., "Effects of the 1981 Depreciation Revisions on the Taxation of Income from Business Capital," National Tax Journal, Vol. 35, 1982, pp. 1-20.

Hanson, Gregory D. and Vernon R. Eidman, "Agricultural Income Tax Expenditures--A Microeconomic Analysis," American Journal of Agricultural Economics, Vol. 67, No. 2, May 1985, pp. 271-278.

Hardesty, Sermin D. and Hoy F. Carman, "Income Tax Simplification Effects on Crop Farm Decision Making," Agricultural Finance Review, Vol. 45, 1985, pp. 11-20.

Harl, Neil E., "Economics of Tax Shelter Activity in the United States," presented at a Seminar on Tax Shelters and Resource Allocation in Agriculture, Iowa State University, Ames, Iowa, August 29, 1985.

Hughes, Dean, W. and Ann Laing Adair, "The Impacts of Recent Changes in Personal Income and Business Profit Taxes on Investments in the Farm Sector," Agricultural Finance Review, Vol. 43, 1983, pp.1-8.

Institute for Contemporary Studies, Federal Tax Reform: Myths and Realities, Michael J. Bosky, Editor, San Francisco, California, 1978.

Lowenberg-DeBoer, James and Michael D. Boehlje, "Analysis of the Impact of the 1981 and 1982 Federal Tax Legislation on Farmers," Research performed for the Economic Research Service, U.S. Department of Agriculture under Research Agreement No. 58-3J23-1-0229X, Iowa State University, July 1984.

Jeremias, Ronald A., James M. Hrubovcak, and Ron L. Durst, "Effective Income Tax Rates for Farm Capital, 1950-1984," National Economics Division, Economic Research Service, U.S. Department of Agriculture, Washington, D.C., June 1983.

Meisner, Joseph C. and V. James Rhodes, "The Changing Structure of U.S. Cattle Feeding," Agricultural Economics Special Report 167, University of Missouri-Columbia, August 1975.

Minarik, Joseph J., Testimony before the Subcommittee on Monetary and Fiscal Policy, Joint Economic Committee, July 1982a.

Minarik, Joseph J., "The Future of the Individual Income Tax," National Tax Journal, Vol. XXXV, No. 3, September 1982b.

Musser, Wesley N., Neil R. Martin, Jr., and Fred B. Sauders, "Impact of Capital Gains Taxation on Farm Organization: Implications for Meat Animal Production on Diversified Farms," Department of Agricultural Economics, University of Georgia, 1976.

Penson, John B. Jr., Robert F.J. Romain, and Dean W. Hughes, "Net Investment in Farm Tractors: An Econometric Analysis," American Journal of Agricultural Economics, Vol. 63, No. 4, November 1981, pp. 629-635.

Reid, D. W., W.N. Musser and N.R. Martin, Jr., "A Study of Farm Firm Growth in the Georgia Piedmont with Emphasis on Intensive Growth in Hog Production," Research Bulletin 249, University of Georgia, College of Agriculture Experiment Station, January 1980.

Simon, W.. E., "Reforming the Income Tax System," American Enterprise Institute Studies on Tax Policy Washington, D.C., and London, 1981.

Sisson, Charles Adair, Tax Burdens in American Agriculture: An Intersectoral Comparison, Iowa State University Press, Ames, Iowa, 1982.

Slemrod, Joel and Shlomo Yitzhaki, "On Choosing a Flat-Rate Income Tax System," National Tax Journal, Vol. XXXVI, No. 1, March 1983, pp. 31-44.

U.S. Department of Agriculture, Agricultural Statistics, 1982.

U.S., Department of Agriculture, Ag Statistics, 1982, Washington, D.C., 1982.

U.S. Department of Commerce, Statistical Abstract of the U.S., 1982.

U.S., Department of Commerce, Bureau of the Census, Statistical Abstract of the United States, 1982-1983, National Data Book and Guide to Sources, 103rd ed., Washington, D.C., 1982.

Vedder, Richard K. and Frenze, C., "Fairness and the Flat Rate Tax," J. of Contemporary Studies 6(1) (Winter 1983).

Vedder, Richard K., "Rich States, Poor States: How High Taxes Inhibit Growth," Journal of Contemporary Studies, Vol. 5, Fall 1982.

Volding, Thomas and Michael Boehlje, "An Economic Evaluation of Cash and Accrual Accounting Methods For Farmers," Staff Paper #50, Department of Economics, Iowa State University, Ames, Iowa, March 1977.

Wickham, Ann B., "Agricultural Leasing: The Lease Versus Buy Decision and Its Implications on Farm Firm Capital Structure," Iowa State University Master's Thesis, Spring 1984.

9

Farm Level Effects of Policy Options and Economic Conditions

Peter J. Barry
and Michael D. Boehlje

An important part of the policy process is understanding the likely actions that various groups or individuals may take in response to policy alternatives, and how these responses might vary with the group's structural characteristics. Sometimes the intended positive effects of a new policy can be offset or countered by unintended or unforeseen actions by the members of these groups. In agriculture, for example, government programs of supply management and price supports generally are intended to stabilize farmers' expectations about commodity prices and incomes, thus reducing the total risk position of their farm businesses. If, however, farmers respond to the reduced risk by borrowing additional funds and expanding farm size, then part or all of the intended policy effect may be lost (Boehlje and Griffin; Gabriel and Baker). Similarly, the prospective liquidity provided by public credit programs may encourage some farmers to accept greater risks in production and marketing than otherwise would occur (see the moral hazard discussion in Chapter 7).

Another important reason for understanding farm level responses to policy options involves the structural characteristics of agriculture. As shown in previous chapters, the wide differences among agricultural production units in such characteristics as business size, tenure position, market coordination, financial leverage, business organization, reliance on off-farm income, and risk attitude and management ability of the farm operators make it difficult to generalize about the economic and financial performance of these units. These differences also provide a complex setting for identifying and evaluating differences in the form and magnitude of responses by these units to changes in public policies and new policy initiatives.

The possible responses by farmers to policy options affecting agricultural finance, and their implications for the structure of agriculture are considered in this chapter. The perspective is a long-term one, describing the responses to the stress conditions of the 1980s, since they will have significant long term effects. We first review some of the structural characteristics of farm businesses and their relationships to the credit policies and programs of the past are reviewed. A conceptual framework for analyzing farm businesses according to commonly used criteria for financial performance is then developed. Following this, the micro implications of policy options will be discussed.

As indicated in Chapter 8, tax policy has a significant and pervasive impact on farm firms, and consequently the structure and organization of agriculture. In fact as noted in that chapter, one of the espoused purposes of tax policy is to explicitly and directly alter firm behavior and consequently the structural characteristics of the industry. Because the discussion of micro impacts of tax policy was provided in detail in Chapter 8, this chapter will focus on the firm level and structural implicaions of financial policy where the impacts are more indirect.

Structural Characteristics of Agriculture

Agricultural production units in the U.S. traditionally have been characterized by relatively small sizes of operation with a largely non-corporate form of business organization that concentrates ownership, management, and risk bearing in the hands of individual farmers and farm families. These size and concentration factors stand in contrast to larger scales of operation in other business sectors in which ownership, management, and labor are specialized functions and risk bearing is spread over numerous corporate shareholders. Exceptions occur among very large farms; they tend to be industrialized specialty operations as in the case of cattle feedlots, poultry units, and orchards. Similarly, some crop farms, hog production units and others have grown to very large sizes with complex arrangements for ownership, management, and financing.

These large scale units are not numerous in number, but they are contributing an increasing proportion of the commercial activity of the agricultural sector. Moreover, many analysts are projecting further growth in the importance of these types of units. This is reflected by the continued movement toward a bimodal distribution of farm sizes in which there are many small, part-time farms, an increasing proportion of large farms, and fewer medium-sized farms.

Despite its small business orientation, agriculture still is considered a capital intensive industry with investments in farmland, buildings, machinery, equipment and breeding livestock dominating the asset structure of most types of farms. This capital intensity has caused much of farmers' total returns to occur as capital gains in land. This was especially evident in the 1970s when both farm income and land values increased sharply. Then, in the 1980s, when farm income weakened and interest rates increased sharply, capital losses on land and other assets occurred.

The use of borrowed funds has also been a major factor in agriculture, especially for larger production units. The seasonality of farm production and the relatively large capital requirements for entering agriculture and building the capacity to generate income are major factors behind the growth in debt capital. In addition, many farmers use credit as a source of liquidity for responding to unanticipated short-falls in farm income. The growth in debt was especially strong in the 1970s, averaging nearly 15 percent per year in the second half of the decade. Asset values grew at comparable rates to debt growth during this period, although the financial risk position of the farm sector increased considerably as well. This, in turn, helped lead to the stress times of the 1980s.

Leasing of farmland has been a wide-spread method of financing that is especially attractive for expanding a farm's scale of operations. Nearly 40 percent of U.S. farms and nearly 65 percent of the farmland have been operated by farmers who lease part or all of the land they operate (U.S. Department of Commerce). Most of the leasing occurs by part owners with a typical pattern of land control characterized by heavy reliance on leasing by younger operators, followed by a combination of ownership and leasing for older farmers, and then by greater reliance on ownership as farmers approach retirement age. Leasing also varies with the type of farm. Full owners have been dominant in fruit and nut, vegetable, and dairy farms. Ownership also is important for most types of livestock operations in which land contributes a lower proportion of the unit's total asset value. Leasing is most common in crop farms where grains, oil seeds, and cotton are the major products.

Leasing of nonreal estate assets has been less extensive in agriculture than in other industries, although it has been increasing (Penson and Duncan; Barry, Hopkin and Baker). Operating leases in agriculture have mostly involved the custom hiring of both machine and labor services for short periods of time. Financial leases mostly have involved machinery, equipment, some buildings, and breeding livestock. Tax considerations have played an important role in financial leases, and beginning in the 1980s both the Cooperative Farm Credit System and some commercial banks began to offer leasing services to farmers along with their lending activities.

In general, then, the financial profile of the farm sector in recent years has been characterized by a relatively strong solvency position due importantly to the appreciation of land values. Chronic liquidity problems and cash flow pressures also have occurred due to the dominance of fixed assets and instabilities in annual returns. At the individual farm level, the effects of these aggregate conditions have been magnified considerably, occasionally causing significant financial problems for many farmers.

In their marketing activities, individual farmers traditionally have had little capacity to influence resource or commodity prices. They have been price takers in both markets. Resource prices generally have experienced upward pressures over time, due to inflation and other factors, while commodity prices have been less responsive. The result is a long-term narrowing of cost-return margins per unit of farm production. Moreover, low elasticities of prices and incomes for many commodities that are subject to weather and other uncontrollable events, including volatile export conditions, have caused wide swings in commodity prices. In response to some of these risks, the vertical linkages and degrees of coordination between agricultural production and other stages of the food and fiber system have continued to evolve over time. The linkages and coordination vary from open markets with commodity prices established at each stage of the system to a vertically integrated system in which successive stages are combined into one decision-making unit.

The degrees and types of vertical coordination vary greatly within agriculture (Sporleder). Nearly all dairy production and most commercial fruit and vegetable production operate under some form of market contract. The poultry industry generally, and broiler production in particular, have moved strongly toward vertical integration. The last decade has witnessed substantial vertical coordination in the beef industry. The swine industry has had similar developments, and even cash grain production has moved away from its strong emphasis on open markets toward greater use of forward contracts by farmers.

In their marketing activities, farmers have sought greater flexibility in timing of sales, more effective use of market information, and improved methods of transferring or sharing risk with other parties. Spreading of sales throughout the year has become more common; so has use of various types of pre-or post-harvest contractual arrangements that increase market coordination with other stages of the food system. While relatively few farmers hedge directly on the futures market, many may reduce price risks indirectly through forward cash contracts for product sales to first handlers, who themselves may hedge on the futures market.

In their production activities, those farm businesses that rely more heavily on land and crops experience a lengthy production process that typically occurs once a year and follows a fixed sequence of activities in producing salable products. This contrasts with the quicker pace of production in other industries in which many of the processes occur simultaneously rather than in sequence. Livestock production and feeding are also subject to lengthy periods and to a fixed sequence of activities, although greater flexibility is provided for scheduling groups or batches of animals. In general, the lengthy sequence of production for most agricultural commodities provides a setting in which farmers have relatively little flexibility in adjusting to changes in policy or other external factors. The production process is largely uncontrollable--once initiated, it seldom ceases, and cropland in particular is seldom held idle unless in response to a government program or as part of a normal rotational pattern.

In general, then, the structural setting within which agricultural producers respond to changes in policy alternatives includes an industry dominated by relatively small businesses with growing concentration of ownership, a highly capital intensive asset structure, considerable use of borrowed money by a large number of farmers, significant leasing of farmland, lengthy production processes that are not conducive to assembly line methods, and a heterogeneous marketing system heavily oriented toward open market transactions. Numerous exceptions occur to these generalizations but, on the whole, they make the agricultural production sector unique relative to other stages of the food and fiber system, and to other industries as well.

Financial Policy and the Structure of Agriculture

Understanding the relationships observed in the past between financial policies and the structure of agriculture is another important factor in evaluating the micro implications of policy options. Since the beginning of this century, a number of significant developments in private markets and public policy have occurred to improve the workings of financial markets for agricultural producers. Some examples of these policy developments include: the creation and evolution of the Cooperative Farm Credit System; the maintenance of a dual system of commercial banking (basically, large and small banks) with some special provisions for financing agriculture; the creation of government credit programs for agriculture--the Farmers Home Administration and Commodity Credit Corporation at the federal level and various credit programs at the state level; the actions and policies taken by the federal and state governments to discourage or impede the flow of outside equity capital into the agricultural production sector; and the encouragement of seller financing of farmland that keeps the financing function within local communities.

Viewed over the long term, readily available credit associated with these types of developments has facilitated various structural changes within the agricultural production sector. Included are the mechanization and modernization of farm units, greater capital intensity, growth in farm size (and reductions in farm numbers), greater leverage from farm debt and leasing, and greater market coordination. Moreover, as indicated earlier, unused credit also plays an important risk bearing role through the liquidity provided to cope with risk and the various alternatives in debt management for restructuring and rescheduling farmers' financial obligations. Besides these positive effects, however, special credit programs and concessionary terms may have highly significant effects on asset values, resource allocation, and risk positions. That is, these credit programs may on occasion tend to over facilitate change, hamper long-term resource adjustments, and yield other adverse effects.

In general, then, the relationships between financing, financial policy, and structural change in agriculture are several fold. First, the availability of credit and other financing methods is a necessary condition for undertaking capital investments and other activities that enhance the well-being of agricultural producers. However, credit availability is not a sufficient condition for improvements since basic profit incentives are needed as well. Second, rural financial markets play an important risk bearing role for agricultural producers and have the potential for adding considerable stability to rural areas and to financial markets in general. Finally, credit and credit policies can be a facilitating instrument for structural change in agriculture. They are not as effective as direct policy instruments, however, since the unintended effects (too much debt, too much risk, resource immobility, and so on) may outweigh the intended positive benefits. Thus, credit markets should not be expected or asked to do too much in accomplishing structural change or in resolving farm income problems.

In like fashion, tax policy may have significant and sometimes unintended implications for individual farms and the structure of agriculture. Because tax policy is so pervasive in management decision-making, it has frequently been chosen as a policy instrument to encourage changes in firm behavior to what are perceived by policymakers as more socially desirable ends. However, the discussion of Chapter 8 clearly indicates that the longer run and aggregate consequences of tax policy may not be consistent with individual short-run behavior or espoused public policy objectives. Consequently, like credit policy, one should not ask too much of tax law as a policy intervention tool. More fundamentally, caution must be exercised in using tax policy to obtain socially desirable goals because the end result in terms of farm structure and financial health of the industry may be exactly the opposite of what was intended.

A Conceptual Framework for Farm Level Effects

In designing a conceptual framework for evaluating the micro implications of policy options, the important components include the goals of the farm operator, the various structural characteristics discussed above, the external events affecting the operating environment of farm businesses, the action choices available to farmers, and the likely effects of these choices on the farmer's goal attainment. Especially important is the set of criteria selected to represent farmers' goals.

Most of the studies of farmers' goals indicate that they, like many other business managers and owners, place considerable emphasis on financial criteria for measuring business performance and evaluating their overall well-being. Especially important are such goals as: 1) reasonable levels of income and growth in net worth, 2) security and stability, and 3) the ability to meet financial obligations. When translated in financial terms these goals are represented by three major criteria for business performance: 1) profitability, 2) risk, and 3) liquidity. Profitability refers to farmers' returns to the equity capital they have invested in their farm businesses. Thus, growth in net worth is a form of profitability measure. Risk refers to possible losses in equity capital and to difficulties in meeting financial obligations due to unanticipated variations in business performance. Liquidity refers to the ability to generate cash in order to meet unanticipated cash demands as they occur. Thus, liquidity is a way of responding to risk, although it is treated here as one of the major performance criteria.

These are conflicting criteria for farmers just as for others. Decision-makers generally prefer high levels of profitability. Risks, however, are to be avoided. Moreover, assets that are more liquid usually are less profitable to hold. Because most people are risk averse, they experience trade-offs among these criteria in achieving the best possible business organization. Simply put, people must take risks and sacrifice liquidity in order to gain profits. Moreover, more risks must be taken and more liquidity given up as more profits are sought. Of course, some risks will materialize and losses will occur, but this is logical to expect. When the risks are high, as in agriculture, then effective risk and liquidity management has a high payoff.

In applying these criteria to farm businesses, it is helpful to use a balance sheet approach in which a structure of assets and liabilities is sought that is "optimal" in terms of a farmer's attitude toward profit and risk. Here, assets and liabilities are defined very broadly. Assets are comprised of all of the items of value that make up the firm. Included is everything the business has control over, whether owned outright or leased. Liabilities are all of the claims on assets and income. Profits then are the returns to the owner's equity capital: the net returns to assets less the costs of financing paid to lenders and lessors. The measure of returns may or may not include unrealized capital gains or losses.

The effects of risk pervade the farmer's balance sheet. The traditional business risks are found on the asset side. They include: 1) production and yield risk 2) market and price risk 3) losses from severe casualties and disasters 4) social and legal risks (changes in tax laws, government programs, trade agreements, and so on) 5) human risks as to the performance of labor and management and 6) risks of technological change and obsolescence. These business risks are distinguished from financial risks that arise on the liability side of the balance sheet. The greater is financial leverage the greater are financial risks in meeting obligations to lenders and lessors. Borrowing risks also come from variations in interest rates and swings in credit availability. Leasing risks come from variations in rental rates and from possible loss of access to leased assets. Thus, like profits, risks are determined by forces affecting both the assets and liabilities of farm businesses. These sources of risk take different forms; they are correlated to varying degrees with one another; and they all bring the threat of financial losses and the promise of financial gain.

The concept of financial leverage is important in agriculture. As used here, leverage refers to the amount of debt capital and other fixed-obligation financing (e.g., leasing) that is used relative to the amount of the firm's equity capital. The principles of finance indicate that increases in financial leverage will increase both the expected level and variability of returns to a firm's equity capital, assuming of course that the rate of return to the assets being financed exceeds the cost of borrowing. Thus, total risk increases as financial leverage increases, and the firm's survival prospects experience greater jeopardy. Optimal leverage for a particular farm business then depends on the farmer's behavioral attitudes toward risk and expected return and on the relationships between the timing of cash flows and repayment obligations. As indicated above, more risk averse farmers may prefer to carry less risk and a lower level of expected return, and thus may have relatively low financial leverage. In contrast, less risk averse farmers typically have higher financial leverage, assuming of course that other farm characteristics are the same. Similarly, the desired level of leverage may differ among farms depending on their size, tenure position, risk management practices, stage in the farmer's life cycle, and other structural and managerial characteristics.

To utilize the balance sheet approach in financial analysis, it is helpful to consider that the individual farmer has determined a target or equilibrium position for the composition of assets and liabilities and their relationships to his equity capital position (Gabriel and Baker; Barry and Baker). Thus, it can be said that the business is in "equilibrium" in terms of the amount of business risk and financial risk being carried, and the liquidity needed for responding to these risks. Within this framework, various changes in the farmer's environment which may influence his equilibrium position can be evaluated along with the effectiveness of possible actions taken to restore equilibrium. These changes might occur as shocks due to the different sources of business risk (crop disasters, unanticipated price changes, swings in land values) and financial risks (higher interest rates), or they might come from new policy initiatives. Whatever the source, these changes will alter the farm's equilibrium position, and provide the incentive for actions to restore equilibrium.

To illustrate, consider again the significant financial stresses affecting many farmers in the 1980s. The reductions in farm income and land values (business risks) along with higher interest rates and high debt loads (financial risks) clearly disturbed the longer term equilibrium position of a number of farmers that had been developed during the boom times of the 1970s. Moreover, the terms of borrowed capital (interest rates, loan maturities, collateral requirements) have changed as well to reflect the risks experienced by lenders in dealing with farmers' financial problems. The increases in both business risk and financial risk clearly have prompted the search for effective managerial and policy responses to allow both farmers and farm lenders to adjust their balance sheets to changing conditions. These responses could take many forms , but they basically focus on reducing the level of business risk, financial risk, or both in farm businesses.

The interrelationships between and within public and private responses are important in this adjustment process. They may have subtle effects that sometimes run counter to the intended objective. Suppose, for example, that a new public policy (e.g., higher commodity price supports, or expanded production insurance) has the effect of substantially reducing the business risk for a group of farmers. If the provisions of the program turned out to be too liberal, then farmers could utilize the program to a point where they overshoot an equilibrium position that is optimal from society's viewpoint. That is, they might respond with actions of their own that are risk increasing such as greater borrowing, faster business expansion, and more venturesome practices in production and marketing. Thus, the intended reduction in business risk from a policy response might be counter balanced by the actions of the decision-makers that increase business risk, financial risk, or both. Clearly, these risk balancing features need careful consideration in the design of public programs and policy instruments.

Responding to Financial Stress in Agriculture

The breadth and depth of the financial problems in agriculture during the 1980s have created a significant dilemma for farmers, lenders, policy makers, and others in how to respond. According to the conceptual framework developed above, the broad choices are clear cut: focus on reductions in business risk and/or financial risk in order to improve the prospects for farm survival and financial viability. In practice, however, the responses that comprise these choices are difficult to implement quickly and effectively, especially when the fundamental causes of low income problems involve pricing and production considerations in domestic and international markets that are beyond the control of individual producers and their lenders. Many of the conventional risk management practices available to farmers are not very effective in resolving serious cash flow problems. For example, most protective responses in a farmer's production and marketing activities are short-term options designed to maintain or stabilize production levels or prices within the production year. They are important for these purposes, but may not help much in dealing with financial crisis situations. Moreover, their routine use may not even be very effective in preventing severe financial stress, if debt loads are relatively heavy and shocks come from the types of external sources cited above.

In response to stress, heavy emphasis has been placed on financial responses to risk, but even here the range of choices is limited. Many farmers who utilized their credit reserves with private sector lenders as a source of liquidity in the early 1980s, when the hard times began, now find that their indebtedness is higher and their remaining borrowing capacity is largely depleted. Lenders have tended to stay with many of their borrowers by exercising forebearance and by utilizing various practices in debt management designed to work out of problem loan situations. Such practices are feasible for interim periods of one to three years, but are not sustainable over long periods. Reductions in land values have eroded away the once stable and growing long-term credit reserves created by equity in farmland. Thus, the capacity to stretch out loan payments through long term refinancing is considerably diminished. Moreover, higher loss rates and costs of administering problem loans have made lenders more cautious in their lending practices, which further diminishes farmers' credit reserves. In addition, these higher lending costs have been passed on to other more successful borrowers through higher interest rates, which has tended to spread the credit problems.

Most financially stressed farmers have responded by slowing down or eliminating capital expenditures, living more frugally, cutting production expenses, emphasizing cash flow relative to profitability, seeking off-farm employment (which was hampered by the general recession of the early 1980s), and selling highly liquid financial assets and inventories of grain and other commodities. As conditions have worsened, the focus shifted more to down-sizing the scale of farm businesses through sales of capital assets including machinery, breeding livestock, and even tracts of land. Some sales were voluntary liquidations on a partial basis; others involved involuntary liquidations and foreclosures. The traditionally thin land market has been stressed even further as more land came available for sale with few if any prospective buyers in some regions. These conditions further pushed down the values of land and other capital assets. While already prominent in many regions, leasing of farmland became even more accepted as a financing mechanism. Cash rents, especially in the Corn Belt, have stayed higher than the dollar values of share rents, reflecting in part the strong demand for leasing. Sale and lease-back arrangements have received greater attention, as have the possibilities for attracting new outside equity capital as a means of relieving high debt loads.

In light of the limited options available to farmers, considerable attention in policy circles has focused on government responses to financial stress in agriculture at both the federal and state levels. The focus here is on the micro implications of these policy choices for farmers. These short-term financial responses are considered in two broad categories: 1) responses affecting the asset structure and income generating activities of farm businesses, and 2) responses affecting the liability structure and financial obligations of farm businesses. It is also important to recognize that many of the suggested policy responses directly involve farm lenders who hold the debt claims on these farm businesses and who are concerned about their own financial well-being.

Policy responses affecting the asset structure of farm businesses generally involve mechanisms to facilitate the liquidation of fixed assets, namely farmland. They are designed to protect the integrity of the asset markets, help buffer the downturn in asset values, and give farmers and their lenders a chance to liquidate or recapitalize farm businesses in an orderly way. A government sponsored entity that is capitalized to purchase farmland from heavily stressed farmers and/or lenders is an example of such a mechanism. The land could be held for a relatively short time (three to five years, or until farm economic conditions improve) and then resold in the market in an orderly fashion, perhaps giving the seller the first purchase option. In the interim, the seller could continue to farm the land under a rental arrangement, with the sale proceeds used to retire farm debt. The seller could be subject to managerial supervision and control in order to qualify for the repurchase option. Numerous variations of this arrangement are possible and, in principle at least, it could be extended to non-land assets.

From an individual farmer's standpoint, this policy response would likely be attractive as a last resort means of preserving a farm business and providing for the recapitalization needed to work toward achieving financial viability in the future. The limited time span, repurchase option, long term leasing arrangement, and perhaps management supervision and control, would all be consistent with the long term goals of building profitability, responding to risk, and restoring liquidity. Similarly, the support provided to the characteristically thin land markets would reduce the financial vulnerability of other farmers who, while not selling land, must still respond to financial problems. Such a program would need careful design, implementation, and administration to assure the greatest prospects for fair and equitable treatment of the parties involved, as well as to avoid extreme effects on the asset markets in different regions and for different farming conditions.

Other policy responses affecting the asset structure and income generating capacity of farm businesses include tax incentives for asset distribution, restraints placed on farmers' costs of production, and increases in commodity prices. Under current tax laws, most farmers report taxable income on a cash basis. Thus, they may not have paid income taxes on a buildup of inventory. Many also face potentially large tax obligations on the sale of capital assets. Finally, the salability of assets by farmers who have built up operating losses could be enhanced if unutilized tax losses could be transferred to new buyers of farm assets. While the financial effects for farmers of such exchanges could be significant, the differences in tax positions among farmers could make such a policy difficult to implement and costly to administer. Moreover, these provisions do not directly aid in preserving the farm unit, only in facilitating its liquidation. These factors, along with the broader fiscal policy issues, make tax changes an unlikely policy response.

Similarly, short-term, interim policies to increase product prices could be ineffective with potentially adverse long-term consequences, especially if export potential in foreign markets is reduced. These actions would not focus directly on credit problems and would run counter to the longer term goals of agricultural policy. Moreover, they would likely be applied on a non-selective basis across all types of farmers and benefit the undeserving as well as those who are competent, but have been adversely affected by actions beyond their own control.

The second category of policy responses involves those affecting the liability structure and financial obligations of farm businesses. These basically address the financial risk component of total risk. Included are those actions affecting the magnitude and cost of debt obligations, the maturity structure of repayment obligations, the infusion of new equity capital, and the financial management skills of farmers. More specifically, actions taken to reduce the level of indebtedness and/or interest rates would reduce the effects of financial leverage and ease the pressure on capital structures. Actions taken to lengthen the maturity structure and stretch out repayment obligations would ease the pressures on near-term cash flows. The infusion of new equity would enable reductions in leverage and a dispersion of risks among a wider set of claimants on farm assets. Finally, programs to improve the financial management skills of farmers could aid in both economic recovery and preventing recurrences in the future.

Under severe stress the quickest and likely most effective response from the farmer's standpoint would be a write-down of indebtedness to levels that are financially feasible to repay over the long term at market interest rates. This approach would not alter interest rates as the basic pricing mechanism, it would leave loan servicing arrangements to the borrower and lender, and it would not change the ownership structure and managerial control of the farm business. Moreover, it would restore the capital structure to pre-stress levels, reduce financial risk, and rebuild liquidity.

Of course, this response would likely be the most costly from the public sector's standpoint and would place no burden of compensation on the farmer. Thus, a closely related approach might involve a conditional loan write-down, perhaps in the form of a loan guarantee, that shares risk among borrowers, lenders, and the government, and assures future compensatory action by farm borrowers. Most people are willing to provide future compensation to cover the cost of benefits received in the near term. Implementing the loan write-down concept in terms of a guarantee or insured loan program enables some of the cost and risk sharing attributes to be included. Moreover, the tasks of credit administration and loan servicing would remain with the farmer's local lender, perhaps with assistance from public credit program personnel. In turn, this contributes to the integrity and discipline needed in successful credit relationships over the long-term.

Under severe financial stress, actions taken to restructure indebtedness and stretch out loan repayments become less favored since they imply a deferring and a pushing ahead of the debt obligation that may then increase rapidly in size. Indeed, liberal debt restructuring in the past may have actually contributed to the current stress problems. Thus, debt restructuring has limited prospects for future success, unless a significant recovery of income generating potential is likely. Alternatively, debt restructuring has greater potential for moderately stressed farms, although here too the prospects for income recovery are important.

Actions taken to infuse new equity capital into farm businesses and disperse the ownership structure likely would meet considerable resistance from farmers. A careful consideration of the costs and benefits of separating the ownership and management of farm assets would be needed. Mechanisms also might be needed to effectively allocate management responsibilities among the equity claimant, to provide for the transfer of equity claims, and to provide for the possible retirement of the equity infusion as business conditions improve. Greater acceptance could occur for equity capitalization involving long-term leasing arrangements on farmland. Many farmers might consider relinquishing outright ownership of farmland and its related financing requirements, if they were assured of the right to use the land in agricultural production over the long term based on acceptable leasing arrangements. This approach combines the restructuring of liabilities with that of farm assets, as discussed above.

Finally, all of the short-term actions could have significant long-term effects on the structure of agriculture. Determining eligibility conditions for farmers to use the programs will determine who remains in farming, who exits from the industry, and the structural characteristics of the units involved. The short-term actions will also determine the viability and vitality of different groups of farmers once economic recovery occurs. In general, there is considerable merit in bringing financial responses more to the forefront of risk management in agriculture. Innovations in loan programs in which farmers' swings in cash flows are buffered on both the down side and the up side need to be considered for the long-term stability they would provide. Such buffering devices would avoid the tendency for excessive risk exposure in boom times and severe stress in periods of adversity.

Longer Term Structural Effects

It also is important to consider a longer term perspective on the farm-level effects of public policies, including those affecting the financial markets. Many of the financial market forces and policy issues were identified and discussed in previous chapters. Here, the farm-level effects of these policy issues and market forces affecting the structure, performance and operating procedures of financial intermediaries are considered.

The changing regulatory and competitive forces in financial markets, including the preference for greater privatization of some credit institutions, will have significant effects on the cost and availability of credit for agriculture. Due to interest rate deregulation, for example, the cost of borrowing for agricultural producers likely will remain higher and more volatile than in pre-1980 times and will follow market interest rates much more closely. Similarly, the continued geographic liberalization of banking regulations and the emergence of more complex financial systems mean that the functions of marketing financial services, loan servicing, and credit decisions will become more distinct. An increasing proportion of credit control and loan authority likely will occur in sub-regional and regional money centers that are located away from the rural areas. This will continue to fragment and dichotomize the farm credit market so that commercial scale agricultural borrowers are treated as part of a financial institution's commercial lending activities (although separate personnel for agricultural and commercial loans should still be prevalent) and smaller, part-time farmers are treated as part of consumer lending programs. This, in turn, will have significant impacts on the structural characteristics of agricultural production units.

Size and Coordination of Production Units

The interest and ability of commercial lenders in financing larger, soundly managed production units will probably continue to increase. This in part reflects the anticipated movement toward larger banking systems as well as toward larger independent banks that will remain active in rural credit markets. Lending capacities of these institutions are increasing due to greater legal lending limits and larger, more stable funding bases. Risk bearing capacities in financial institutions are also increasing as a result of larger sizes and expanded geographic scope, and from greater diversity in lending activities. These factors will encourage individual banks and banking systems to seek the business of larger farm units and agri-businesses that are considered the more profitable borrowing customers. Greater efficiency and cost control in customer relationships will be sought with large loan volumes continuing to play a role in lending efficiency. Similarly, as the Farm Credit System continues to evolve into a full service, united, competitive, commercial financial institution, it is natural for its banks and associations to seek the financing of larger, well managed farm businesses. This may increase the financing costs and reduce the credit availability for higher risk borrowers, including young farmers, limited resource farmers, and part-time operations. Entry into farming will, therefore, become difficult and greater concentration can be expected.

The evolution of the financial system toward larger scale, more open financial markets will put more emphasis on a systems approach to financing agriculture. The systems approach involves the vertical linkages and degrees of coordination that occur among the various stages of the food and fiber system. The linkages and coordination may vary from largely open markets with prices established for commodities at each stage of the system to a vertically integrated system in which the successive stages are combined into one decision-making unit. For individual farmers, the systems approach involves a more formal, comprehensive consideration of the combined effects of production, capital investments, financing, and marketing in their business planning. Marketing plans will need to be soundly developed in terms of the type of practices typically followed for the various commodities. The awareness and use of forward contracting, hedging, spreading sales, and monitoring market information need to move into a more mature status in business planning with rigorous, yet routine use for the various commodities involved.

From the lenders' standpoint, the systems approach

makes financing less risky, more efficient, easier to control, and more profitable. The systems approach to agricultural financing has been especially common in large money center banks, especially those on the West Coast, and in some regional banks. As these types of banks and related banking systems continue to develop national lending markets and engage in joint financing of larger agricultural units, the systems approach to credit analysis and financing will become more prominent. Thus, agricultural borrowers will be commonly expected to consider the combined effects of production, capital investments, and marketing in their financial planning, and to have soundly developed business plans. The systems approach does not necessarily imply a movement toward greater market coordination, although this will likely occur for many commodities. A systems approach does, however, imply greater attention by borrowers to integrating the financial, production, and marketing components of their management activities.

Financial and Tenure Structure of Farms

The farm credit problems of the mid-1980s have hampered a longer term assessment of farmers' financial responses to the regulatory and competitive changes affecting financial institutions. It is clear that part of the effects of financial deregulation (higher, more volatile interest rates) and the related responses of financial institutions (e.g., floating loan rates) have contributed to the increased leverage and financial vulnerability of these commercial scale producers. Over the long-term, however, the financial structure issue in agriculture rests heavily on the balance between 1) the levels and variabilities of returns to assets for agricultural production units with different structural characteristics, and 2) the levels and variabilities of interest rates and other costs of financial capital.

If the levels and volatility of borrowing costs remain higher relative to farm asset returns as a result of financial deregulation, or if rates of return remain lower and more volatile, then agricultural borrowers must take actions to reduce the financial leverage in their operations or seek other methods of responding to the higher financial risks such as hedging and other marketing strategies, diversification, and holding more financial assets. Between these two choices, the leverage response will likely be the greater, especially for those farms under greater financial stress. Lenders in general prefer levels of farm debt that can be serviced with reasonable confidence from projected cash flows; in the absence of sharp increases in farm revenues, this preference means a reduction in leverage relative to current levels.

Other leverage pressures are arising from the recent declines in land values that have reduced equity levels, diminished long term credit reserves, eroded refinancing capacity, and brought greater uncertainties about future changes in land values. The continued high variabilities in both farm income and interest rates suggest that greater variability of land values likely will occur in the future. This is a significant change from the steady, dependable up trend in land values, and thus refinancing capacity, that characterized agriculture from World War II until the late 1970s. As a result, the risks associated with future levels of farm wealth, long-term credit reserves, and refinancing capacity will be much greater than in the past. In addition the importance of liquid assets in the overall asset structure of agricultural production units will likely increase as well, to cope with a volatile future environment. Thus, fundamental changes in the financial structure of agriculture will occur as a result of the higher costs and wider swings of interest rates in financial markets. While the long-term boom and bust cycles in agriculture will continue to occur, these market and regulatory factors suggest that they could be more pronounced than in the past, unless more effective protective and stabilization methods are adopted.

The future may also bring changes in both farmers' use and the availability of credit from various sources, although this will depend on the evolving competitive and structural characteristics of financial institutions. For commercial banks, the greater lending competition among larger banks and banking systems, along with a shift of final credit decisions away from rural communities, may make agricultural borrowers more sensitive to changes in interest rates and other financing terms. The traditionally strong lender-borrower relationship at the local level may diminish with more shopping for credit occurring, especially among small and medium sized farms. An offsetting factor, however, could be the growing competition by larger banks for the more profitable agricultural customers through bank marketing practices and personalized services. The ability of commercial banks to develop stable, cost competitive sources of long-term credit will also stabilize their relationships with the more profitable borrowers. In general, however, as banking systems further develop, the profitability and risk characteristics of agricultural lending will be scrutinized more closely in terms of how they compare with other types of lending. Farm lending likely will remain an effective source of diversification in loan portfolios unless low profitability and high risks become chronic conditions.

The pattern of financing could also differ for borrowers from the Farm Credit System in light of the System's trend toward unifying the management and administration of its short-, intermediate-, and long-term lending programs. The same loan officer serving the full range of these programs may encourage more borrowers to obtain complete financing from the Farm Credit System, compared to split financing of the past in which the same borrower might utilize a Federal Land Bank for long-term lending and perhaps commercial banks or other sources for short-and intermediate-term loans. Greater uniformity in risk assessment and credit arrangements should occur, as well as a stronger emphasis on the borrower's financial management.

The future may also bring changes in the equity capital structure of agriculture with continued efforts to develop channels for attracting outside equity. For the traditional owner-operator farm unit, the major sources of equity capital have been retained farm earnings, capital gains especially on farmland, off-farm earnings, and gifts and inheritances from other family members. Over time, however, intergenerational transfers have dispersed the ownership of farmland so that off-farm landlords and others have become significant equity claimants. Moreover, some subsectors of agriculture (cattle feeding, cattle breeding, citrus and vineyard) have departed substantially from the small business, family orientation by placing considerable reliance on outside equity capital. In some cases, the equity investments have responded to tax considerations and short-term profit prospects. Considerable instability in such funding can occur if capital is withdrawn for reinvestment elsewhere due to changes in tax laws, profit prospects, or other factors.

In the future, the opportunities for attracting outside equity into agriculture involve the public attitude toward this funding source and the needs for new funding methods. The needs for outside equity largely are based on the financing required for high cost capital items and the spreading of business and financial risks over a deeper equity base and a more diverse set of investors. Neither of these needs for equity capital likely will diminish. The cost of establishing and operating farm businesses will continue to increase, and business and financial risks will remain relatively high. In addition, the greater sophistication in management for utilizing new technologies may also lead to greater separation and specialization in the financing, ownership, and management functions. Thus, the regulatory and competitive conditions affecting farm credit markets, combined with greater sophistication in managerial skills, will continue to stimulate financial innovations involving outside equity capital.

Leasing has been a primary method of financing the control of farmland by farm operators, and is receiving greater use for nonreal estate assets. In general, leasing of any asset is a form of financial leveraging since the lease creates a fixed financial obligation (rent) for the lessee, as well as a valuable asset in the lease contract. Within this framework, leases may exhibit a variety of characteristics. In many cases, tax considerations may strongly influence the leasing terms, while market forces, local customs, and economic conditions of the leasing parties also are important.

Financial institutions have shown growing interest in offering financial leases to farmers at lower costs than intermediate-term loans. The financial institutions can do this through the tax savings they realize from owning the leased assets. These developments suggest that financial leasing may continue to evolve into a routine method of financing that is used by some producers but not others, and the use of which will vary with tax and other economic factors.

In looking to the future and considering the

changing regulatory and competitive environment of financial insitutions, it is likely that longer term leasing will become more common in agriculture. Compared to smaller banks, larger banks and banking systems are more accustomed to offering leasing services along with credit and financial services to their customers. In addition, the Farm Credit System also has begun to offer financial leases to their customers, and many merchants and dealers also offer leasing arrangements.

It is more difficult to project whether leasing of farmland will increase; it is already extensive. However, significant changes in leasing practices may occur. For example, lenders and other financial institutions likely will become less concerned with the method of resource control (ownership, lease), and more concerned with the stability and cost of resource control. Thus, financial institutions will be increasingly responsive to leasing contracts that assure the asset's availability and use for long periods of time. Similarly, financial plans by agricultural producers that document the cost advantages of leasing will become more important, as will the producer's skills in managing leasing arrangements with multiple landlords and other lessors.

Concluding Comments

This chapter has focussed on the important relationships among financial policies for agriculture, the likely impacts and responses by farmers, and the long- term structural characteristics of agriculture. After reviewing the current structural characteristics, a conceptual framework for evaluating the farm level implications of various policy options was developed. This framework emphasized the important interrelationships among a farm's financial structure, its risk position, and a farmer's behavioral attitudes toward risk and expected return. In the process it was shown how increasing levels of business risk could be offset by reductions in financial risk, or vice versa. Also demonstrated was the high sensitivity of farmers' responses to various policy options affecting their risk position.

The conceptual framework was then used to evaluate

how various policy options would affect farmers who are experiencing considerable financial stress in the 1980s. Responses affecting both the asset and liability structures of farm businesses were considered. For the asset structure, the policy responses generally involve mechanisms to facilitate the orderly liquidation of fixed assets, especially farmland. For the liability structure, the policy responses mostly involve actions affecting the magnitude and cost of debt obligations, the maturity structure of repayment obligations, the infusion of new equity, and the improvement of the financial management skills of farmers. An important point was the emphasis on short-term transition policies that avoided the deferring and piling up of financial obligations in the future, unless the prospects for income recovery were significant.

The chapter was concluded by considering the long-term structural effects of public policies, current stress conditions, and the evolving forces in financial markets. Especially important were the prospects for higher and more volatile farm interest rates, as well as the geographic restructuring and consolidations affecting both commercial banking and the Farm Credit System. The consequences for agriculture include a continuing push toward larger sizes of farm businesses; a separation of commercial lending for commercial scale farm businesses and consumer-oriented lending for smaller, part-time farms; greater emphasis on integrating the production, marketing, and financial components of farm businesses; greater consideration of innovative methods of leasing and equity capital financing in agriculture; and more formal management of financial leverage and credit by farmers. All these factors combine to present special challenges for farmers, lenders, policy-makers, and others to choose appropriate strategies and policies for restoring financial health and vitality to the agricultural sector.

REFERENCES

Barry, P. J., "Regulatory and Performance Issues for Financial Institutions: Their Effects on Technology Adoption and Structural Change in Agriculture," Office of Technology Assessment, U.S. Congress, Washington, D.C., 1986.

Barry, P. J. and C. B. Baker, "Financial Responses to Risk" in Risk Management in Agriculture, P. J. Barry, Editor, Iowa State Univeristy Press, Ames, Iowa, 1984.

Barry, P. J., J. C. Hopkin, and C. B. Baker, Financial Management in Agriculture, 3rd ed. Interstate Printers and Publishers, Danville, Il., 1983.

Boehlje, M. D. and J. R. Brake, "Solutions (or Resolutions) of Financial Stress Problems from the Private and Public Sectors," American Journal of Agricultural Economics, 67(1985): forthcoming.

Boehlje, M. D. and V. Eidman, "Financial Stress in Agriculture: Implications for Producers," American Journal of Agricultural Economics, 65(1983): 937-944.

Boehlje, M. D. and S. Griffin, "Financial Impacts of Government Price Support Programs," American Journal of Agricultural Economics, 61(1979): 285-296.

Gabriel, S. C. and C. B. Baker, "Concepts of Business and Financial Risk" American Journal of Agricultural Economics, 62(1980): 560-564.

Lee, W. F., M.D. Boehlje, A. Nelson, and W. Murray, Agricultural Finance, 7th edition, Iowa State University Press, Ames, Iowa, 1980.

Penson, J. B. and M. Duncan, "Farmers Alternatives to Debt Financing," Agricultural Finance Review, 41(1981): 83-91.

Sporleder, T. L., "Emerging Information Technologies and Agricultural Structure" American Journal of Agricultural Economics 65(1983): 388-395.

U.S. Department of Commerce, Bureau of the Census, 1979 Farm Finance Survey, 1978 Census of Agriculture, July, 1982.

10

Summary and Conclusions

Dean W. Hughes

Financial information plays an important role in farm policy-making. The income statement and the balance sheet of the farm sector have been estimated for over forty years, far longer than similar data have been available for other parts of the economy. The successes and failures of most farm policies are judged on the basis of their impacts on these measures of financial well-being. Direct involvement in the capital markets serving farmers has been an important part of government involvement in agriculture throughout this century. Special tax rules for farmers have also been important in determining the growth in agricultural output through their influence on the after-tax returns in farming versus other industries.

The extent of the currently available data on financial conditions in the farming sector was described in Chapters 2 and 3. Historic trends that have led to current problems include several factors, but all have their basis in a substantial decline in farm income after an increase in the early 1970s. Along with depressed levels of farm income in the 1980s have come higher interest rates and a substantial increase in the variability, or risk, of farm earnings. The increased fluctuations in income have been magnified by the growing leverage of the sector's balance sheet, making the risks of investing in the sector grow dramatically over the last 15 years. Less stable incomes can be explained by the integration of agriculture into the rest of the economy through an increased reliance on international trade and the deregulation of financial markets. The bearing of risk is normally compensated by higher returns. Given the declines in incomes of the 1980s, the only way for rates of return to increase has been for asset values to decline. As a result, farmers' wealth, collateral values and credit reserves have been reduced. Such adjustments have caused significant problems for many farmers and others in rural areas.

Disaggregating national measures of financial condition by size of farm, by region of the country and by types of commodities produced help to identify what segments of the farm population are experiencing the most problems, as shown in Chapter 3. Medium-sized farms are least able to withstand current difficulties since large farms are more profitable and small farms have the greatest off-farm income. Using changes in debt-to-asset ratios as indicators of the deterioration of financial conditions, the Corn Belt and the Northern Plains regions of the country have experienced the most rapid decline in wealth and liquidity since 1980. Disaggregating national income statements and balance sheets by type of commodity supports the regional evidence regarding areas of financial stress. Financial statements for cash grain farms showed the most problems. And, these farms are concentrated in the Corn Belt and the Northern Plains.

Chapter 4 presented a description of the traditional sources of equity and debt capital for farmers. Given the prevailing ownership patterns of farms, growth in equity has been far more difficult to achieve than changes in debt. Most farm equity comes from unrealized capital gains on farm assets and retained earnings. While there are many available avenues for the generation of outside equity for farm firms, there is little evidence that any have been used extensively. Farm debt comes from both private and government lenders. The Farm Credit System, which is owned by its member-borrowers, is the largest lender to the sector. A residual category of lenders, called Individuals and Others, holds the second largest amount of farm debt. Commercial banks, while the largest lender of nonreal estate debt, are the third largest lenders overall. Finally, insurance companies have played a significant, though declining, role in farm real estate markets. The two principal federal government organizations that lend directly to farmers are the Farmers Home Administration and the Commodity Credit Corporation.

Discussions of historic information and potential future policy choices were linked together in Chapter 5 through a discussion of how financial projections for the sector are made. Recognizing the integration of the farm sector with the rest of the economy, projections of future financial conditions were presented under three alternative macroeconomic policy regimes. The projections suggest continued problems for U.S. primary industries, including farming, unless or until the Federal Government is able to reduce its borrowing needs to cover large deficits. Much of the discussion of policy options in later chapters was predicated on expectations of continuing problems for farmers.

Chapter 6 dealt with the current and future issues in private farm capital markets including debt, equity and leasing. Clearly, decisions that have been made and will be made in regulating banking behavior will influence the cost and availability of credit in the farm sector. These changes will influence the size and structure of farms. Deregulation of commercial banks has already lead to fewer credit shortages, but also to higher and more volatile interest rates for the farm sector. It has increased competition among banks, lowered their profit margins and generated stresses for those banks facing losses on their agricultural loans. The Farm Credit System has also gone through some regulatory changes in the last five years, and questions continue to be raised regarding the System's connections to the Federal Government. Many difficult choices will have to be made by the System and the government in the coming years as the implications of farm financial difficulties are realized by the largest lender to the sector. Since equity capital represents ownership interests, public policy choices regarding farm equity markets are at the heart of determining the future structure of the sector. The historic reluctance of the American public to allow separation of ownership from management of farms will be tested as the sector struggles with the inability to meet the fixed debt commitments during the current period of financial adjustment. Leasing provides a third method of obtaining the use of an asset. The advantages of leasing change with the changes in tax laws and other financial regulations, and policymakers' decisions will be important in determining how important leasing is to farmers in the future.

The reasons for direct government involvement in farm credit markets, specific issues related to the implementation of government credit programs and a description of current financial policy options were presented in Chapter 7. Justification for government intervention in a market is based primarily on a failure of the market to distribute resources in a way deemed best by society. Government lending to farmers, particularly that of the Farmers Home Administration, has been justified on the grounds that certain individuals who cannot get credit from private lenders could, with an initial loan from the government, establish a viable farming operation and later graduate to private credit. In some of their programs, the Farmers Home Administration has been able to satisfy a legitimate public need. In others, however, questions have been raised about the appropriateness of government involvement due to direct and indirect impacts of such programs.

Currently available short-run responses to financial stress in the farm sector include use of the bankruptcy laws, debt moratoria, loan guarantees, debt and asset restructuring of balance sheets, interest rate reduction plans, changes in macroeconomic policies, the development of a government agency to postpone sales of illiquid assets held by lenders, and recapitalization schemes. Most of these proposals would yield some relief to financially stressed farmers and their lenders. They all have costs, however, and their basic differences relate to when and who will bear them. Over the long-run a movement away from most direct government involvement in financial markets is probably appropriate, given the integration of the domestic and international credit markets.

Chapter 8 describes how tax policy interacts with finance to help determine the size and structure of farm firms and the farm sector. Several types of taxes and their effects on agriculture are discussed, including: income, estate, employment, property and sales taxes. Also included are discussions of the importance of tax law changes as determined by empirical studies. Generally, tax laws have had a major role in determining asset values, investment decisions, the production of specific agricultural commodities, ownership patterns and the structure of the farm sector. Policymakers are likely to make choices on the use of cash accounting by farmers, changes in depreciation patterns, tax shelter plans, property tax changes due to declining land values, taxation of cooperative earnings, flat tax rates and tax rate indexing to inflation, which when combined with the financial policy choices described in Chapters 6 and 7, will have major impacts on the farm sector.

Finally, Chapter 9 described the likely reactions to changes in financial policy at the individual farm level. First, a brief description of some of the typical characteristics of U.S. farms was presented along with a framework to understand how returns can be balanced against different types of risks in farmers' investment decisions. The framework was then used to describe the longer run effects of some of the short-run policies for responding to farmers' financial problems in the 1980s. A description of some of the likely farm level responses to the new economic environment for farmers was presented. Included were: the continued move to a bimodal agriculture as small farm loans are treated as consumer loans with large farm loans being sought after by lenders as business loans; significantly improved integration of financial, production and marketing plans and thus more sophisticated ownership and management practices; and greater use of new leasing and equity arrangements.

U.S. farmers and ranchers are working through a financial adjustment that is unprecedented in the lives of almost everyone involved. The costs of the adjustment are many and significant. Consequently, several solutions to such problems are being debated. Those solutions that seem to overcome difficulties with a simplistic approach may be considered appealing by many of the individuals involved.

Hopefully this book has made at least three points: (1) the U.S. farm sector is <u>not</u> a homogeneous blend of medium-sized farms, each growing a variety of farm commodities and being financed in the same way; (2) the rural financial markets are now almost completely integrated into the national and international economies; and (3) the obvious short-run implications of particular financial policies can be substantially different from long-run impacts.

Since farming is not composed of many similar units, the financial problems faced by a particular sized operation producing a specific commodity in one location are not the same as any other farmer's. Broad policies that do not differentiate among farmers, overlook these differences and may misallocate the intended aid. Such misallocations can bring about unintended consequences and add significantly to the costs of a policy response. Moreover, policies that might be good for some farmers could, at the same time, hurt others. Methods are required to weigh these costs and benefits in ways that are meaningful to the farm sector and to society.

The integration of financial markets serving farmers with the rest of the economy has several important implications. Farm problems may not have farm causes or solutions. It is entirely possible that the current "overcapacity" of the sector is a direct result of macroeconomic policies, which have decreased both foreign and domestic demands for U.S. farm production and increased interest expenses. Financial policy solutions to such problems may end up generating additional adjustment costs in the future when macroeconomic policies are changed. Moreover, many of the historic justifications for the existence of farm financial policies have been outdated with the breakdown of the isolation of rural economies. That is not to say that good reasons for farm programs do not exist, but with new justifications come new objectives, and at least adjustments, in old programs. Finally, interactions between the farm and nonfarm economies have grown, and more sophistication is required in designing public policies to help farmers.

If the implications of short-run financial policies are not the same as the long-run impacts, then even more care needs to be given to financial policy formulation. The benefits and costs of different policies need to be tested over multiple years. Combinations of policies should be sought that are politically palatable in the short-run and lead the sector toward long-run goals. Furthermore, unbiased models need to be developed and agreed upon in order to consistently test the implications of different policy directions.

All three of these points suggest that the simple, easily understood and intuitively appealing solutions to farm financial problems require close scrutiny before acceptance. In fact, it can be argued that simple solutions to current problems are the least likely to be appropriate over the long-run. At a minimum, those who are suggesting, advocating or deciding upon such policies should carefully consider these positions from several different perspectives and be aware of all the data and analytical capacity currently available.

Index

About the Authors

DEAN W. HUGHES is an Associate Professor and Director of the Thornton Agricultural Finance Institute in the Department of Agricultural Economics at Texas Tech University. He has been acting section leader of the Agricultural Finance Research Section of the U.S. Department of Agriculture and an economist with the Federal Reserve Bank of Kansas City. He has an A.B. degree from Princeton Unviersity, a M.S. degree from Purdue University and a Ph.D. from Texas A & M University. He has published articles and made presentations in the areas of agricultural finance, agricultural sector linkages to the national economy, and multisector macroeconomic modelling.

STEPHEN C. GABRIEL is general partner of the consulting firm Farm Sector Economics Associates, a Washington, D. C. based firm specializing in analyzing farm sector financial conditions and macroeconomic linkages to agriculture. He is also editor for agricultural finance for the firm's monthly report, Farm Financial Conditions Review. He previously worked at the Economic Research Service at the U.S. Department of Agriculture where he conducted research and analysis in the area of agricultural finance and credit. He earned a B.A. degree in economics at Loyola University of Chicago. At the University of Illinois he received an M.S. in finance and a Ph.D. in agricultural economics.

PETER J. BARRY is Professor of Agricultural Finance at the University of Illinois. His undergraduate and graduate degrees also came from the University of Illinois. His professional experience includes service in the Departments of Agricultural Economics at the University of Guelph (Ontario, Canada) and at Texas A & M University. His professional activities include research and teaching in agricultural finance with emphasis on financial management, risk management, and financial markets for agriculture. He participates in workshops and conferences with agricultural lenders, policy groups, farmers, and others. His publications include professional journal articles, books, bulletins, and magazine articles.

MICHAEL D. BOEHLJE is Head of the Department of Agricultural and Applied Economics at the University of Minnesota. Prior to moving to Minnesota he was a Professor of Economics and Assistant Dean of the College of Agriculture at Iowa State University. He was on the faculty at Oklahoma State University prior to his Iowa State appointment. Dr. Boehlje received a B.S. degree at Iowa State University and an M.S. and Ph.D. at Purdue University. Dr. Boehlje has taught undergraduate and graduate courses in agricultural finance, holds extension programs in finance with farmers and lenders, is a frequent speaker at lenders' conferences, schools and workshops and is also involved in research in agricultural finance and farm management. Dr. Boehlje is the co-author of two other books, Agricultural Finance, 7th edition, Iowa State University, Ames, Iowa, 1980, and Farm Management, John Wiley & Sons, Inc., New York, 1984. He is also the author or co-author of over 200 articles and publications.

For Product Safety Concerns and Information please contact our EU representative GPSR@taylorandfrancis.com Taylor & Francis Verlag GmbH, Kaufingerstraße 24, 80331 München, Germany

Batch number: 10725895

Printed by Printforce, the Netherlands